Introduction to the Brief Atlas of the Human Body

During my many years of teaching anatomy and physiology, I have found that supplemental images of body structures often help students appreciate the "big picture" of human anatomy and physiology. For that reason, I have provided some of these images in a handy supplement to *Anatomy & Physiology* to help you with your learning.

This *Brief Atlas* contains images from some of the world's best medical atlases. It provides a manageable overview of the human body in an easy-to-use size and format. You will find it helpful for learning in both the lecture/discussion part of the course and in the laboratory part of your anatomy and physiology course.

Part 1, *Surface Anatomy* provides a brief overview of the surface structures of the body. Notice that many of these images feature the locations of major structures that lie just under the skin. These underlying structures include bones, muscles, and various other internal organs. These images will help you understand the relationship of the internal organs to the surface view of the body.

Part 2, *Skeleton* features a number of detailed photographs of the human skeleton. These images include many views of the different regions of the human skeleton and of individual bones of the skeleton. You will find these to be useful references as you study the bones and muscles of the body.

Part 3, *Internal Anatomy* provides a survey of helpful images of the inside structures of the body. These images often include representations of dissected organs of many different systems to help you see how they all fit together in the human body. I have also included some casts, images produced by filling up hollow body structures (such as blood vessels) with a substance that hardens to produce a molded casting of what the hollow spaces look like. Such casts help you appreciate body structure in a different way than ordinary anatomical images.

Part 4, *Cross-Sectional Anatomy* features an introductory set of horizontal sections of the human body. Such sectional views help you develop a stronger sense of the three-dimensional relationships of the various structures of the body. Because of the increasing importance of cross-sectional anatomy in today's medical imaging, I believe that this resource is especially useful in helping you apply your understanding of human body structure in practical, applied ways.

Part 5, *Histology* is a mini-atlas of some of the major types of tissues found in the body. My experience as a student and teacher tells me that an understanding of tissues, or histology, is a good foundation for easily learning the anatomy of every part of the body. My experience also tells me that the more examples you see early in your studies, the easier it is to grasp the fundamental concepts of histology. So this section of the *Brief Atlas* provides a handy set of tissue samples gleaned from the major histology atlases to supplement those already found in *Anatomy & Physiology*.

I trust that you will find these carefully selected images helpful in your study of human anatomy and physiology!

Kevin T. Patton

Brief Atlas and Quick Guide

for Anatomy & Physiology Tenth Edition

Kevin T. Patton, PhD

BRIEF ATLAS OF THE HUMAN BODY, p. 1

- Surface anatomy of the human body
- Photos of dissected organs and bodies
- Photo atlas of bones of the human skeleton
- Cross sections of the human body
- Casts of human organs
- Histology specimens

QUICK GUIDE TO THE LANGUAGE OF SCIENCE AND MEDICINE, p. 91

- Hints for learning and using scientific terms
- Word parts commonly used as prefixes, suffixes, and roots
- Abbreviations used for anatomical directions
- Eponyms and their equivalents
- Scientific, medical, and chemical acronyms, abbreviations, and symbols
- Greek alphabet and Roman numerals

ELSEVIER

ELSEVIER

3251 Riverport Lane
St. Louis, Missouri 63043

BRIEF ATLAS AND QUICK GUIDE FOR
ANATOMY & PHYSIOLOGY, TENTH EDITION

ISBN: 978-0-323-52902-0

Previous editions copyrighted 2016, 2013, 2010, and 2007.

International Standard Book Number: 978-0-323-52902-0

Executive Content Strategist: Kellie White
Content Development Manager: Lisa Newton
Content Development Specialist: Melissa Rawe
Publishing Services Manager: Julie Eddy
Book Production Specialist: Clay S. Broeker
Design Direction: Brian Salisbury

Printed in Canada

Last digit is the print number: 9 8 7 6 5 4 3 2 1

Contents

Surface Anatomy

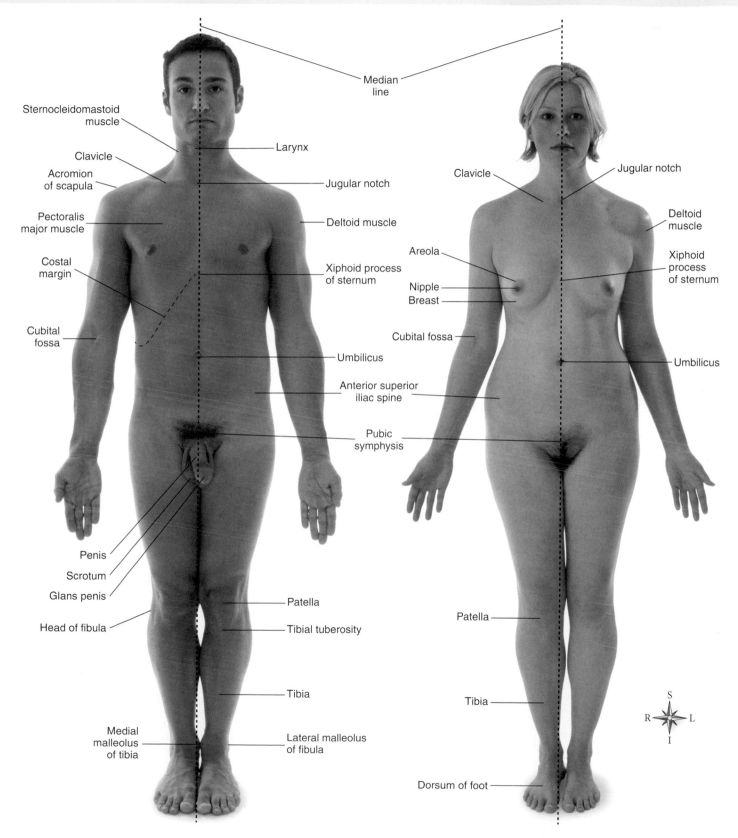

Figure 1-1 *Overview of surface anatomy (anterior view).* *(From Drake RL et al: Gray's atlas of anatomy, ed 2, Philadelphia, 2015, Elsevier.)*

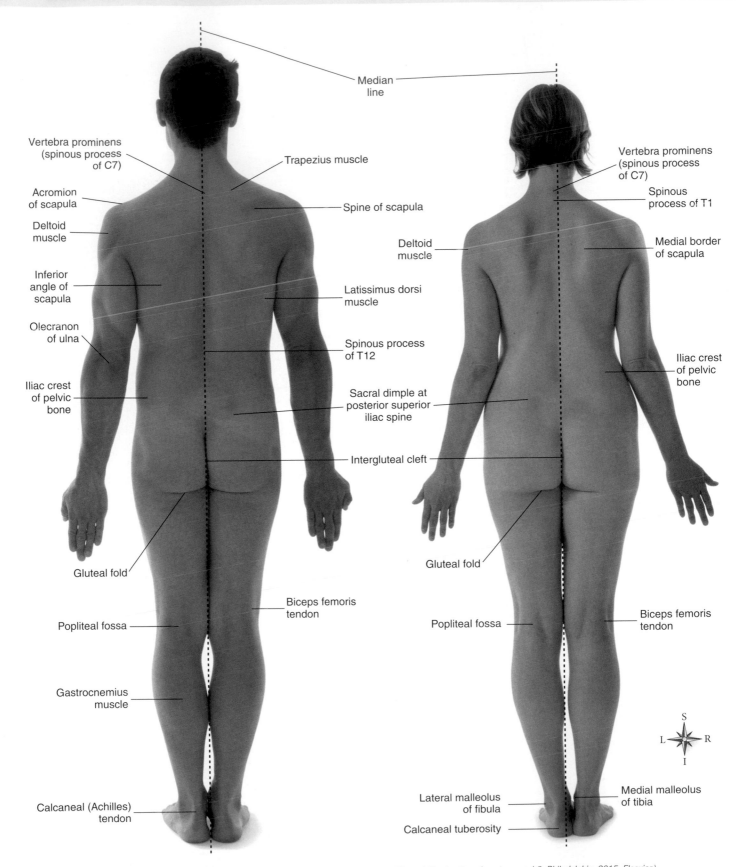

Figure 1-2 *Overview of surface anatomy (posterior view). (From Drake RL et al:* Gray's atlas of anatomy, *ed 2, Philadelphia, 2015, Elsevier.)*

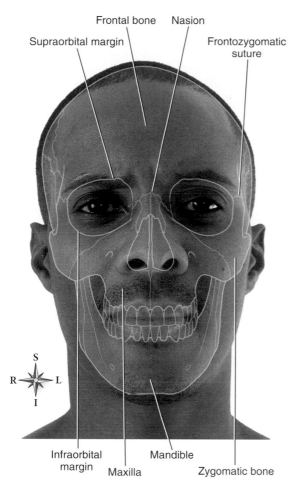

Frontal bone   Nasion
Supraorbital margin   Frontozygomatic suture

Infraorbital margin   Mandible
Maxilla   Zygomatic bone

Figure 1-3 *Head (anterior view).* *(From Drake RL et al: Gray's atlas of anatomy, ed 2, Philadelphia, 2015, Elsevier.)*

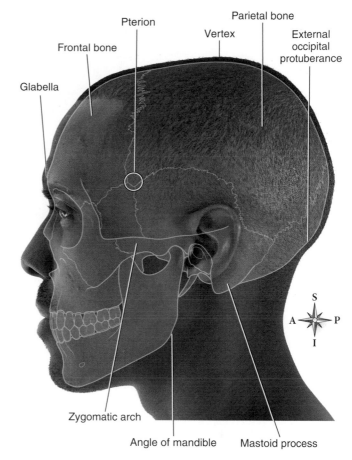

Pterion   Parietal bone
Frontal bone   Vertex   External occipital protuberance
Glabella

Zygomatic arch

Angle of mandible   Mastoid process

Figure 1-4 *Head (lateral view).* *(From Drake RL et al: Gray's atlas of anatomy, ed 2, Philadelphia, 2015, Elsevier.)*

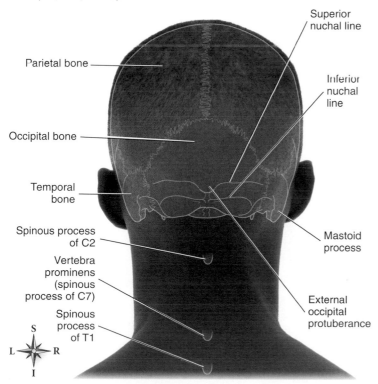

Superior nuchal line
Parietal bone
Inferior nuchal line
Occipital bone
Temporal bone
Spinous process of C2
Mastoid process
Vertebra prominens (spinous process of C7)
Spinous process of T1
External occipital protuberance

Figure 1-5 *Head (posterior view).* *(From Drake RL et al: Gray's atlas of anatomy, ed 2, Philadelphia, 2015, Elsevier.)*

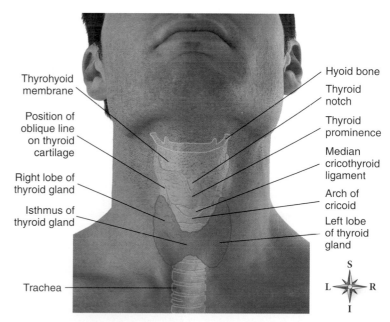

Thyrohyoid membrane   Hyoid bone
Thyroid notch
Position of oblique line on thyroid cartilage   Thyroid prominence
Median cricothyroid ligament
Right lobe of thyroid gland   Arch of cricoid
Isthmus of thyroid gland   Left lobe of thyroid gland
Trachea

Figure 1-6 *Neck (anterior view).* *(From Drake RL et al: Gray's atlas of anatomy, ed 2, Philadelphia, 2015, Elsevier.)*

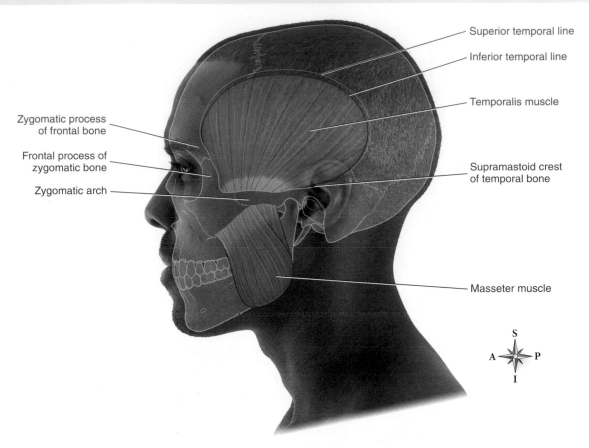

Figure 1-7 *Head and neck (left lateral view).* *(From Drake RL et al: Gray's atlas of anatomy, ed 2, Philadelphia, 2015, Elsevier.)*

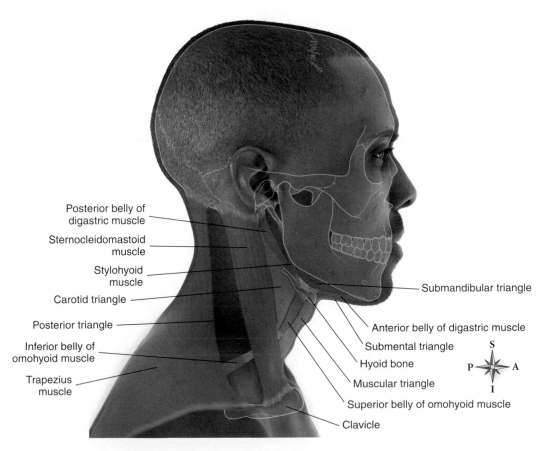

Figure 1-8 *Head and neck (right lateral view).* *(From Drake RL et al: Gray's atlas of anatomy, ed 2, Philadelphia, 2015, Elsevier.)*

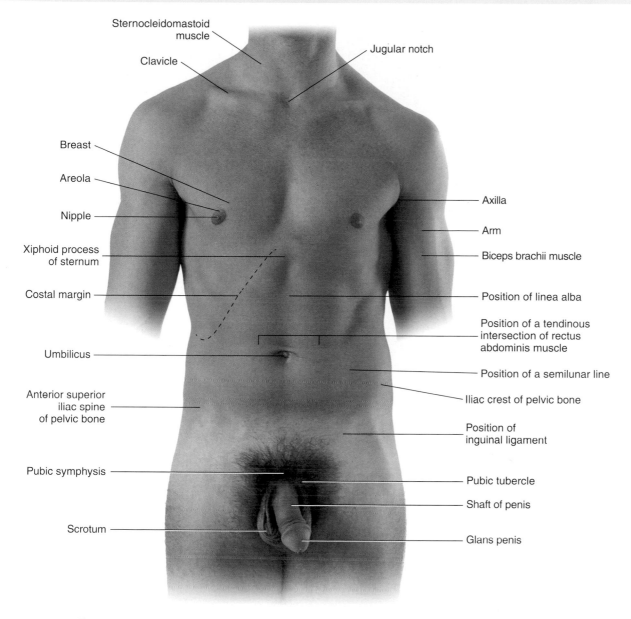

Sternocleidomastoid muscle

Clavicle

Jugular notch

Breast

Areola

Nipple

Xiphoid process of sternum

Costal margin

Umbilicus

Anterior superior iliac spine of pelvic bone

Pubic symphysis

Scrotum

Axilla

Arm

Biceps brachii muscle

Position of linea alba

Position of a tendinous intersection of rectus abdominis muscle

Position of a semilunar line

Iliac crest of pelvic bone

Position of inguinal ligament

Pubic tubercle

Shaft of penis

Glans penis

Figure 1-9 *Male trunk (anterior view).* *(From Drake RL et al: Gray's atlas of anatomy, ed 2, Philadelphia, 2015, Elsevier.)*

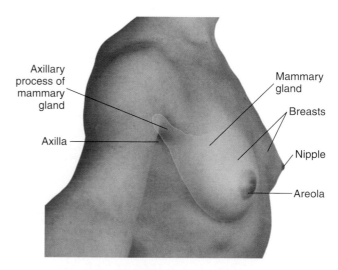

Axillary process of mammary gland

Axilla

Mammary gland

Breasts

Nipple

Areola

Figure 1-10 *Female thorax (right lateral view).* *(From Drake RL et al: Gray's atlas of anatomy, ed 2, Philadelphia, 2015, Elsevier.)*

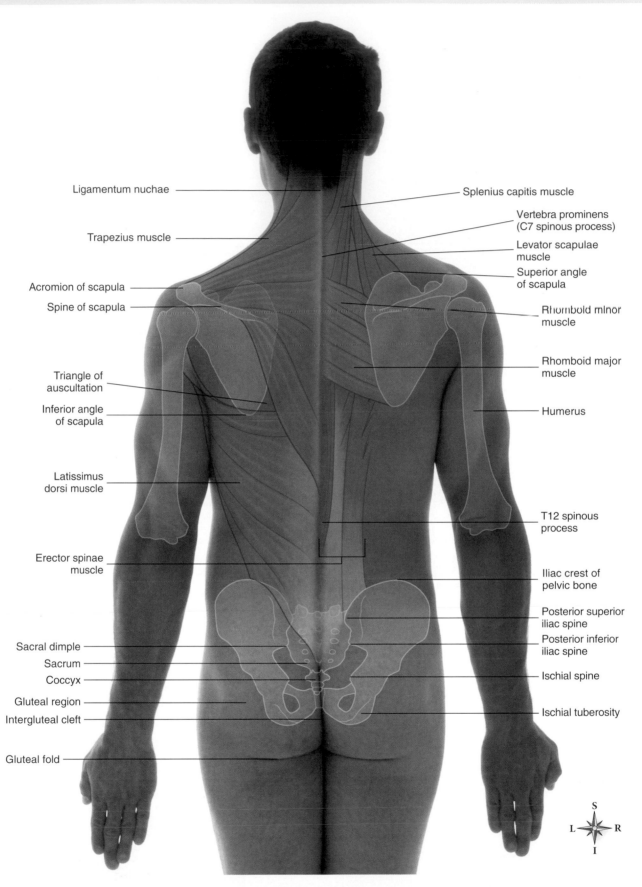

Ligamentum nuchae

Trapezius muscle

Acromion of scapula

Spine of scapula

Triangle of auscultation

Inferior angle of scapula

Latissimus dorsi muscle

Erector spinae muscle

Sacral dimple

Sacrum

Coccyx

Gluteal region

Intergluteal cleft

Gluteal fold

Splenius capitis muscle

Vertebra prominens (C7 spinous process)

Levator scapulae muscle

Superior angle of scapula

Rhomboid minor muscle

Rhomboid major muscle

Humerus

T12 spinous process

Iliac crest of pelvic bone

Posterior superior iliac spine

Posterior inferior iliac spine

Ischial spine

Ischial tuberosity

S
L ✦ R
I

Figure 1-11 *Surface of trunk (posterior view).* *(From Drake RL et al:* Gray's atlas of anatomy, *ed 2, Philadelphia, 2015, Elsevier.)*

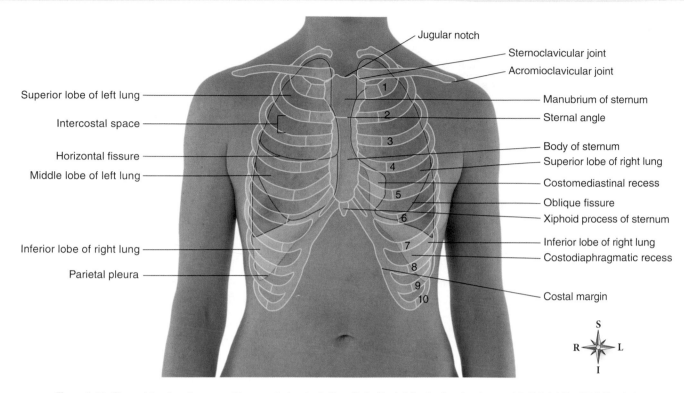

Jugular notch

Sternoclavicular joint

Acromioclavicular joint

Superior lobe of left lung

Intercostal space

Horizontal fissure

Middle lobe of left lung

Manubrium of sternum

Sternal angle

Body of sternum

Superior lobe of right lung

Costomediastinal recess

Oblique fissure

Xiphoid process of sternum

Inferior lobe of right lung

Inferior lobe of right lung

Costodiaphragmatic recess

Parietal pleura

Costal margin

Figure 1-12 *Thorax (showing rib cage and lungs; anterior view).* *(From Drake RL et al:* Gray's atlas of anatomy, *ed 2, Philadelphia, 2015, Elsevier.)*

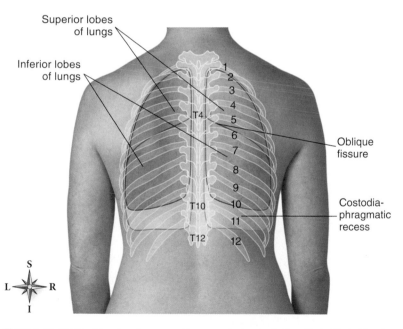

Superior lobes of lungs

Inferior lobes of lungs

Oblique fissure

Costodia-phragmatic recess

Figure 1-13 *Thorax (showing rib cage and lungs; posterior view).* *(From Drake RL et al:* Gray's atlas of anatomy, *ed 2, Philadelphia, 2015, Elsevier.)*

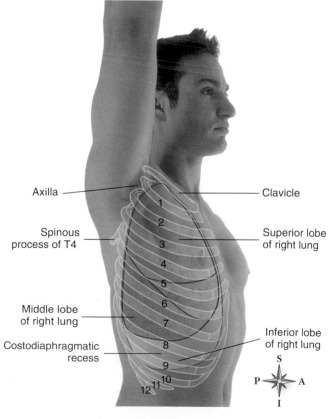

Axilla

Clavicle

Spinous process of T4

Superior lobe of right lung

Middle lobe of right lung

Inferior lobe of right lung

Costodiaphragmatic recess

Figure 1-14 *Thorax (showing rib cage and lungs; right lateral view).* *(From Drake RL et al:* Gray's atlas of anatomy, *ed 2, Philadelphia, 2015, Elsevier.)*

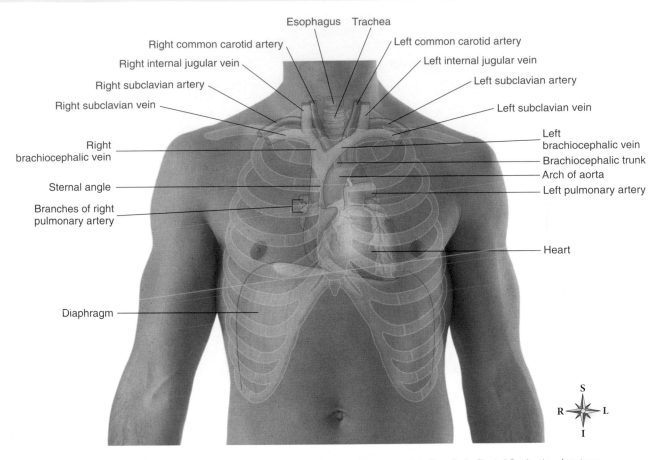

Figure 1-15 *Thorax (showing structures of the mediastinum; heart and large vessels).* *(From Drake RL et al: Gray's atlas of anatomy, ed 2, Philadelphia, 2015, Elsevier.)*

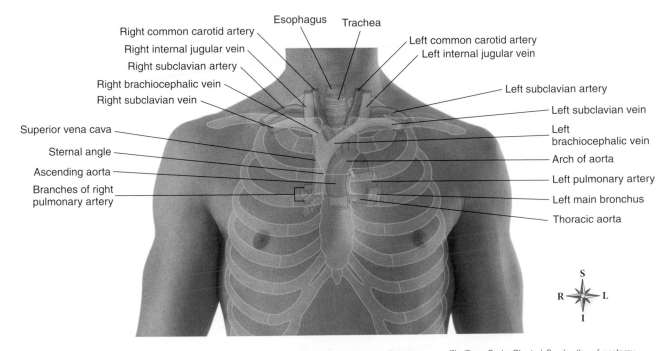

Figure 1-16 *Thorax (showing structures of the mediastinum; large vessels and airways [heart removed]).* *(From Drake RL et al: Gray's atlas of anatomy, ed 2, Philadelphia, 2015, Elsevier.)*

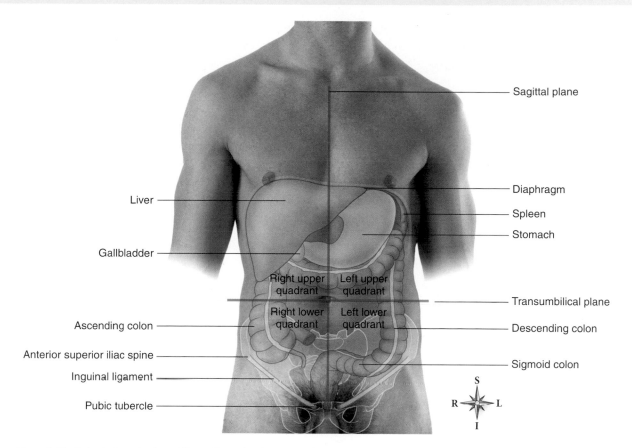

Figure 1-17 *Abdominal quadrants and the positions of major organs. (From Drake RL et al: Gray's atlas of anatomy, ed 2, Philadelphia, 2015, Elsevier.)*

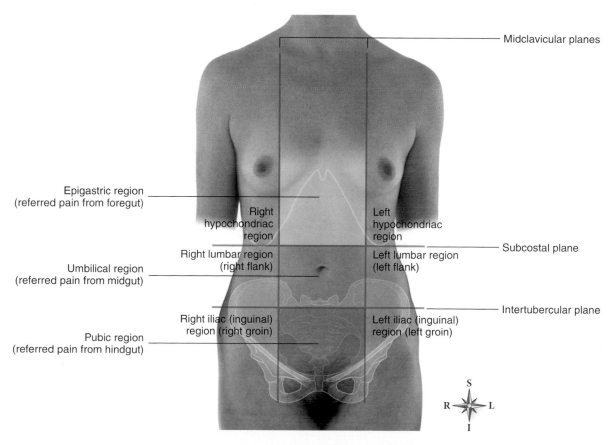

Figure 1-18 *Nine regions of the abdomen. (From Drake RL et al: Gray's atlas of anatomy, ed 2, Philadelphia, 2015, Elsevier.)*

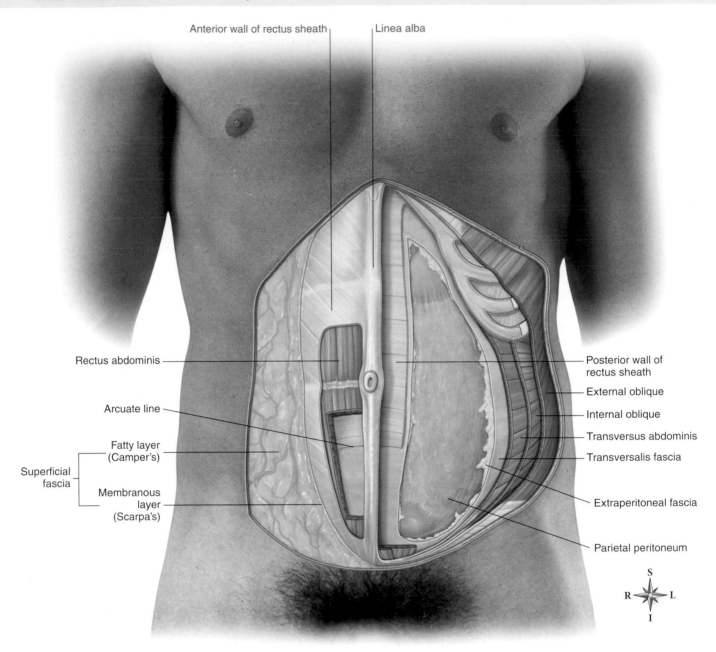

Anterior wall of rectus sheath   Linea alba

Rectus abdominis

Arcuate line

Fatty layer
(Camper's)

Superficial
fascia

Membranous
layer
(Scarpa's)

Posterior wall of
rectus sheath

External oblique

Internal oblique

Transversus abdominis

Transversalis fascia

Extraperitoneal fascia

Parietal peritoneum

Figure 1-19 *Layers of the abdominal wall. (From Drake RL et al:* Gray's atlas of anatomy, *ed 2, Philadelphia, 2015, Elsevier.)*

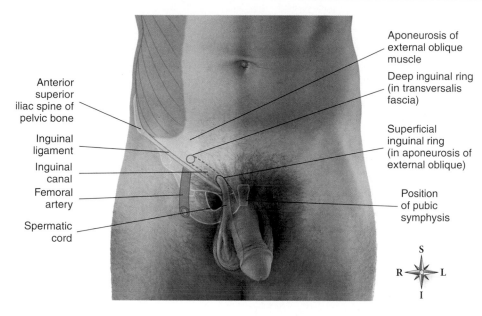

Aponeurosis of external oblique muscle

Deep inguinal ring (in transversalis fascia)

Superficial inguinal ring (in aponeurosis of external oblique)

Position of pubic symphysis

Anterior superior iliac spine of pelvic bone

Inguinal ligament

Inguinal canal

Femoral artery

Spermatic cord

Figure 1-20 *Inguinal region (male). (From Drake RL et al:* Gray's atlas of anatomy, *ed 2, Philadelphia, 2015, Elsevier.)*

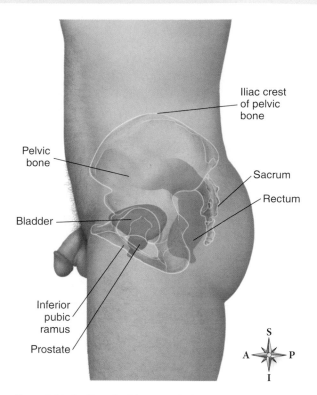

Iliac crest of pelvic bone

Pelvic bone

Sacrum

Rectum

Bladder

Inferior pubic ramus

Prostate

Figure 1-21 *Position of pelvic organs (male). (From Drake RL et al:* Gray's atlas of anatomy, *ed 2, Philadelphia, 2015, Elsevier.)*

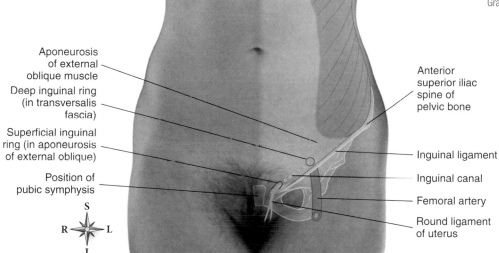

Aponeurosis of external oblique muscle

Deep inguinal ring (in transversalis fascia)

Superficial inguinal ring (in aponeurosis of external oblique)

Position of pubic symphysis

Anterior superior iliac spine of pelvic bone

Inguinal ligament

Inguinal canal

Femoral artery

Round ligament of uterus

Figure 1-22 *Inguinal region (female). (From Drake RL et al:* Gray's atlas of anatomy, *ed 2, Philadelphia, 2015, Elsevier.)*

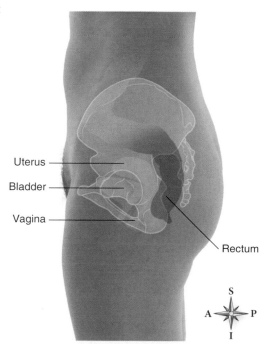

Uterus

Bladder

Vagina

Rectum

Figure 1-23 *Position of pelvic organs (female). (From Drake RL et al:* Gray's atlas of anatomy, *ed 2, Philadelphia, 2015, Elsevier.)*

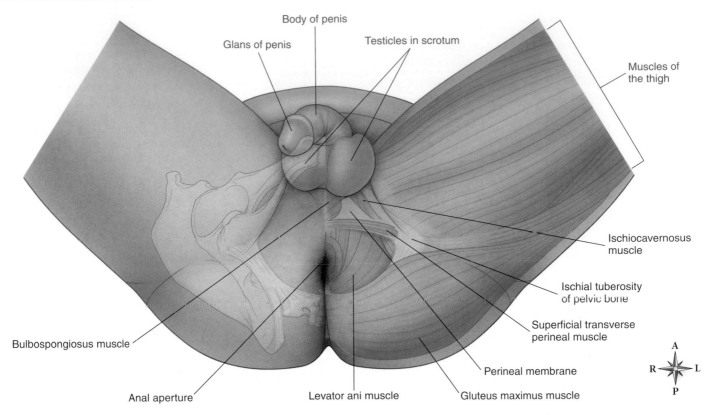

Body of penis

Glans of penis

Testicles in scrotum

Muscles of the thigh

Ischiocavernosus muscle

Ischial tuberosity of pelvic bone

Superficial transverse perineal muscle

Bulbospongiosus muscle

Perineal membrane

Anal aperture

Levator ani muscle

Gluteus maximus muscle

Figure 1-24 *Perineum (male). (From Drake RL et al: Gray's atlas of anatomy, ed 2, Philadelphia, 2015, Elsevier.)*

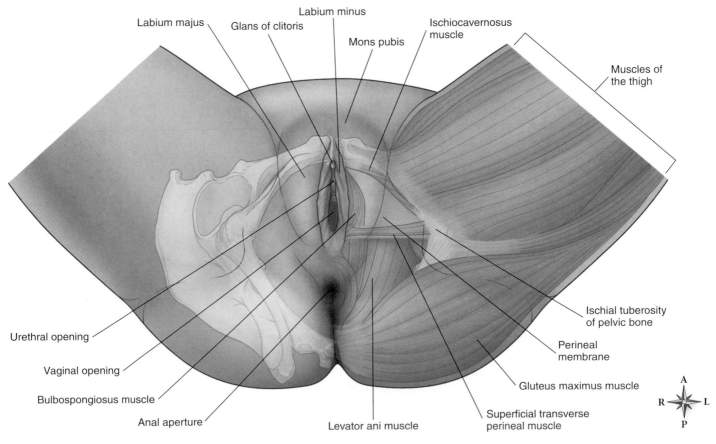

Labium minus

Labium majus

Glans of clitoris

Mons pubis

Ischiocavernosus muscle

Muscles of the thigh

Ischial tuberosity of pelvic bone

Perineal membrane

Urethral opening

Vaginal opening

Gluteus maximus muscle

Bulbospongiosus muscle

Anal aperture

Levator ani muscle

Superficial transverse perineal muscle

Figure 1-25 *Perineum (female). (From Drake RL et al: Gray's atlas of anatomy, ed 2, Philadelphia, 2015, Elsevier.)*

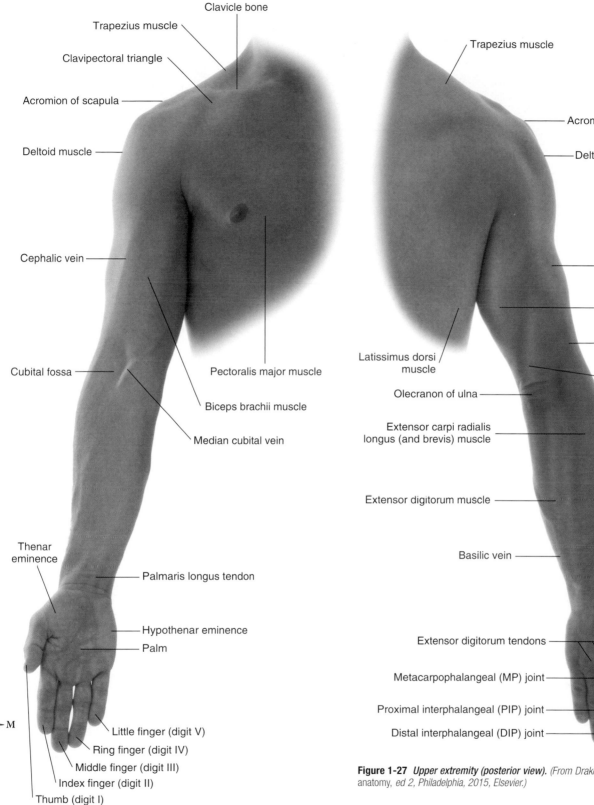

Clavicle bone

Trapezius muscle

Clavipectoral triangle

Acromion of scapula

Deltoid muscle

Cephalic vein

Cubital fossa

Pectoralis major muscle

Biceps brachii muscle

Median cubital vein

Thenar eminence

Palmaris longus tendon

Hypothenar eminence

Palm

Little finger (digit V)

Ring finger (digit IV)

Middle finger (digit III)

Index finger (digit II)

Thumb (digit I)

Trapezius muscle

Acromion of scapula

Deltoid muscle

Triceps brachii muscle (lateral head)

Triceps brachii (long head)

Brachioradialis muscle

Latissimus dorsi muscle

Triceps brachii tendon

Olecranon of ulna

Extensor carpi radialis longus (and brevis) muscle

Extensor digitorum muscle

Basilic vein

Extensor pollicis longus tendon

Extensor digitorum tendons

Metacarpophalangeal (MP) joint

Proximal interphalangeal (PIP) joint

Distal interphalangeal (DIP) joint

Figure 1-26 *Upper extremity (anterior view).* *(From Drake RL et al: Gray's atlas of anatomy, ed 2, Philadelphia, 2015, Elsevier.)*

Figure 1-27 *Upper extremity (posterior view).* *(From Drake RL et al: Gray's atlas of anatomy, ed 2, Philadelphia, 2015, Elsevier.)*

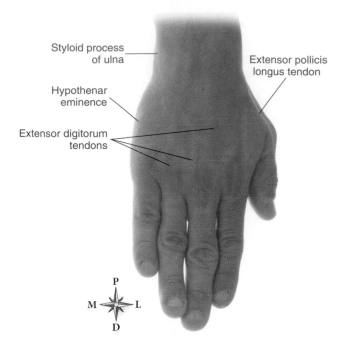

Styloid process of ulna

Hypothenar eminence

Extensor digitorum tendons

Extensor pollicis longus tendon

P
M · L
D

Figure 1-28 *Hand (dorsal surface).* (From Drake RL et al: Gray's atlas of anatomy, *ed 2, Philadelphia, 2015, Elsevier.)*

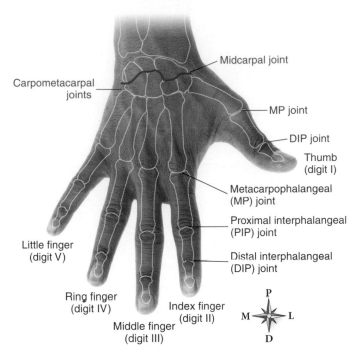

Carpometacarpal joints

Midcarpal joint

MP joint

DIP joint

Thumb (digit I)

Metacarpophalangeal (MP) joint

Proximal interphalangeal (PIP) joint

Distal interphalangeal (DIP) joint

Little finger (digit V)

Ring finger (digit IV)

Middle finger (digit III)

Index finger (digit II)

P
M · L
D

Figure 1-29 *Hand (position of internal structures).* (From Drake RL et al: Gray's atlas of anatomy, *ed 2, Philadelphia, 2015, Elsevier.)*

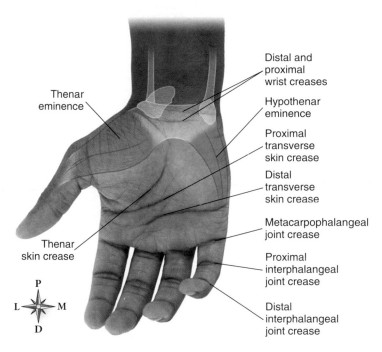

Thenar eminence

Thenar skin crease

Distal and proximal wrist creases

Hypothenar eminence

Proximal transverse skin crease

Distal transverse skin crease

Metacarpophalangeal joint crease

Proximal interphalangeal joint crease

Distal interphalangeal joint crease

P
L · M
D

Figure 1-30 *Hand (palmar surface).* (From Drake RL et al: Gray's atlas of anatomy, *ed 2, Philadelphia, 2015, Elsevier.)*

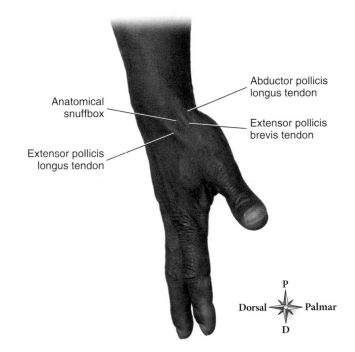

Anatomical snuffbox

Abductor pollicis longus tendon

Extensor pollicis brevis tendon

Extensor pollicis longus tendon

P
Dorsal · Palmar
D

Figure 1-31 *Hand (lateral surface).* (From Drake RL et al: Gray's atlas of anatomy, *ed 2, Philadelphia, 2015, Elsevier.)*

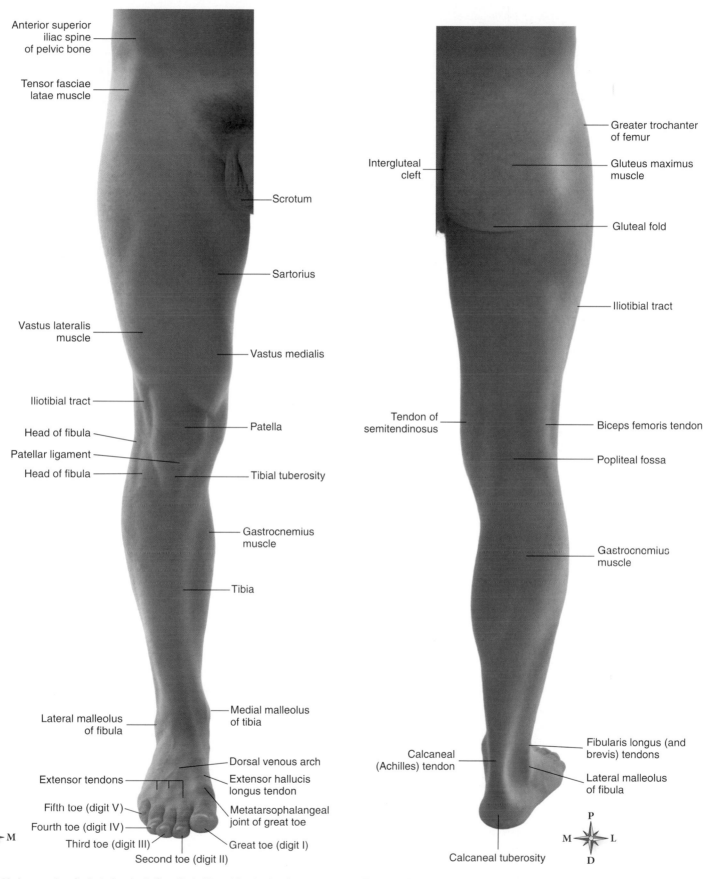

Anterior superior iliac spine of pelvic bone

Tensor fasciae latae muscle

Scrotum

Sartorius

Vastus lateralis muscle

Vastus medialis

Iliotibial tract

Patella

Head of fibula

Patellar ligament

Head of fibula

Tibial tuberosity

Gastrocnemius muscle

Tibia

Lateral malleolus of fibula

Medial malleolus of tibia

Dorsal venous arch

Extensor tendons

Extensor hallucis longus tendon

Fifth toe (digit V)

Fourth toe (digit IV)

Third toe (digit III)

Second toe (digit II)

Metatarsophalangeal joint of great toe

Great toe (digit I)

P
L — M
D

Greater trochanter of femur

Intergluteal cleft

Gluteus maximus muscle

Gluteal fold

Iliotibial tract

Tendon of semitendinosus

Biceps femoris tendon

Popliteal fossa

Gastrocnemius muscle

Calcaneal (Achilles) tendon

Fibularis longus (and brevis) tendons

Lateral malleolus of fibula

Calcaneal tuberosity

P
M — L
D

Figure 1-32 *Lower extremity (anterior view).* *(From Drake RL et al: Gray's atlas of anatomy, ed 2, Philadelphia, 2015, Elsevier.)*

Figure 1-33 *Lower extremity (posterior view).* *(From Drake RL et al: Gray's atlas of anatomy, ed 2, Philadelphia, 2015, Elsevier.)*

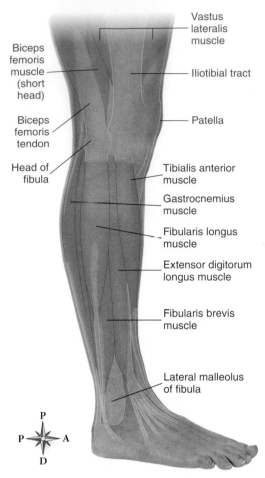

Vastus
lateralis
muscle

Biceps
femoris
muscle
(short
head)

Iliotibial tract

Biceps
femoris
tendon

Patella

Head of
fibula

Tibialis anterior
muscle

Gastrocnemius
muscle

Fibularis longus
muscle

Extensor digitorum
longus muscle

Fibularis brevis
muscle

Lateral malleolus
of fibula

P

P ← → A

D

Figure 1-34 *Muscles of the leg (surface projection; right lateral view).* *(From Drake RL et al: Gray's atlas of anatomy, ed 2, Philadelphia, 2015, Elsevier.)*

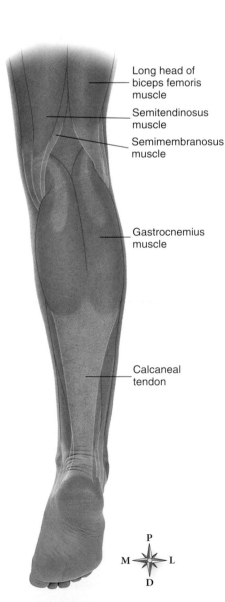

Long head of
biceps femoris
muscle

Semitendinosus
muscle

Semimembranosus
muscle

Gastrocnemius
muscle

Calcaneal
tendon

P

M ← → L

D

Figure 1-36 *Muscles of the leg (surface projection; posterior view).* *(From Drake RL et al: Gray's atlas of anatomy, ed 2, Philadelphia, 2015, Elsevier.)*

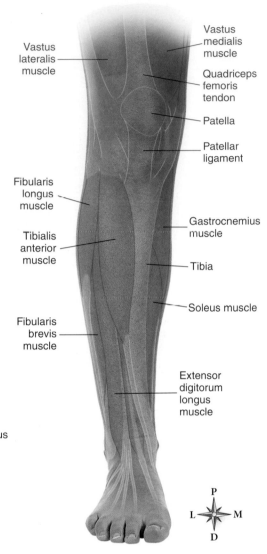

Vastus
lateralis
muscle

Vastus
medialis
muscle

Quadriceps
femoris
tendon

Patella

Patellar
ligament

Fibularis
longus
muscle

Gastrocnemius
muscle

Tibialis
anterior
muscle

Tibia

Soleus muscle

Fibularis
brevis
muscle

Extensor
digitorum
longus
muscle

P

L ← → M

D

Figure 1-35 *Muscles of the leg (surface projection; anterior view).* *(From Drake RL et al: Gray's atlas of anatomy, ed 2, Philadelphia, 2015, Elsevier.)*

Skeleton

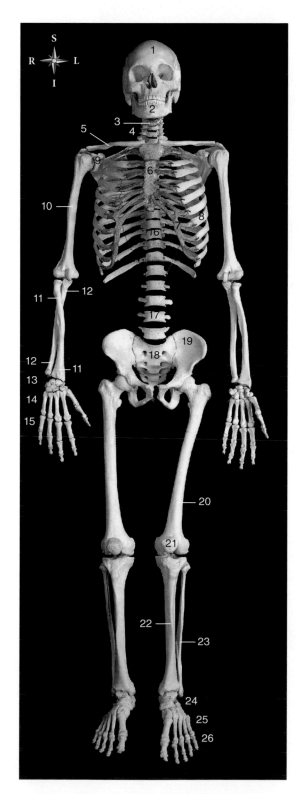

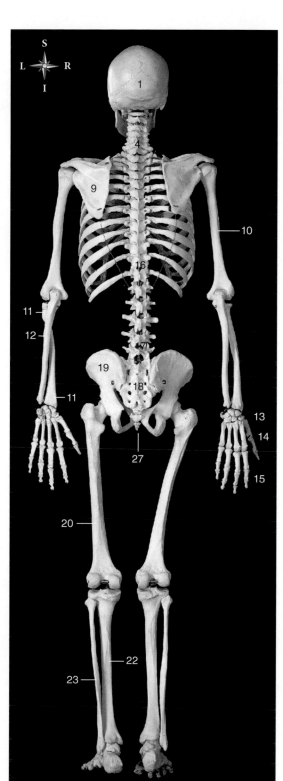

1 Skull
2 Mandible
3 Hyoid bone
4 Cervical vertebrae
5 Clavicle
6 Sternum
7 Costal cartilages
8 Ribs
9 Scapula
10 Humerus
11 Radius
12 Ulna
13 Carpal bones
14 Metacarpal bones
15 Phalanges of thumb
   and fingers
16 Thoracic vertebrae
17 Lumbar vertebrae
18 Sacrum
19 Hip bone
20 Femur
21 Patella
22 Tibia
23 Fibula
24 Tarsal bones
25 Metatarsal bones
26 Phalanges of toes
27 Coccyx

Figure 2-1 *Skeleton (anterior view).*

Figure 2-2 *Skeleton (posterior view).* The left forearm is supinated and the right forearm is pronated.

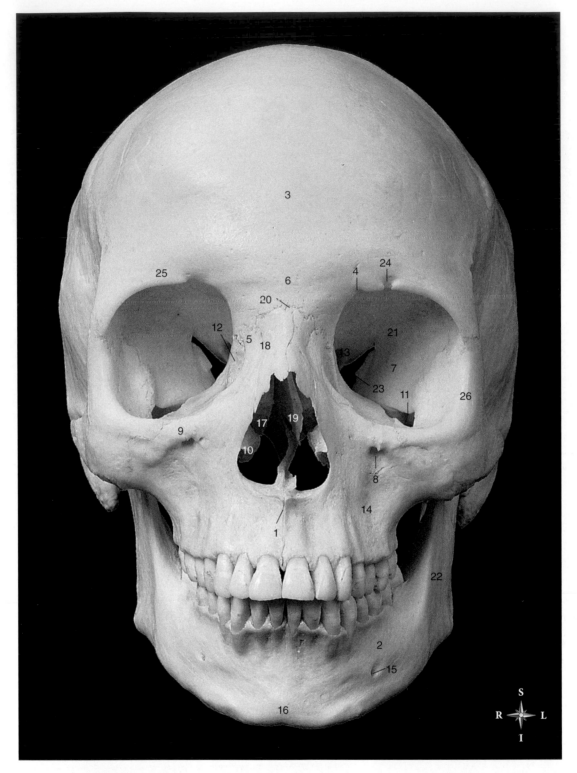

1 Anterior nasal spine
2 Body of mandible
3 Frontal bone
4 Frontal notch
5 Frontal process of maxilla
6 Glabella
7 Greater wing of sphenoid
   bone
8 Infraorbital foramen
9 Infraorbital margin

10 Inferior nasal concha
11 Inferior orbital fissure
12 Lacrimal bone
13 Lesser wing of sphenoid
    bone
14 Maxilla
15 Mental foramen
16 Mental protuberance
17 Middle nasal concha
18 Nasal bone

19 Nasal septum
20 Nasion
21 Orbit (orbital cavity)
22 Ramus of mandible
23 Superior orbital fissure
24 Supraorbital foramen
25 Supraorbital margin
26 Zygomatic bone

Figure 2-3 *Skull (frontal view).*

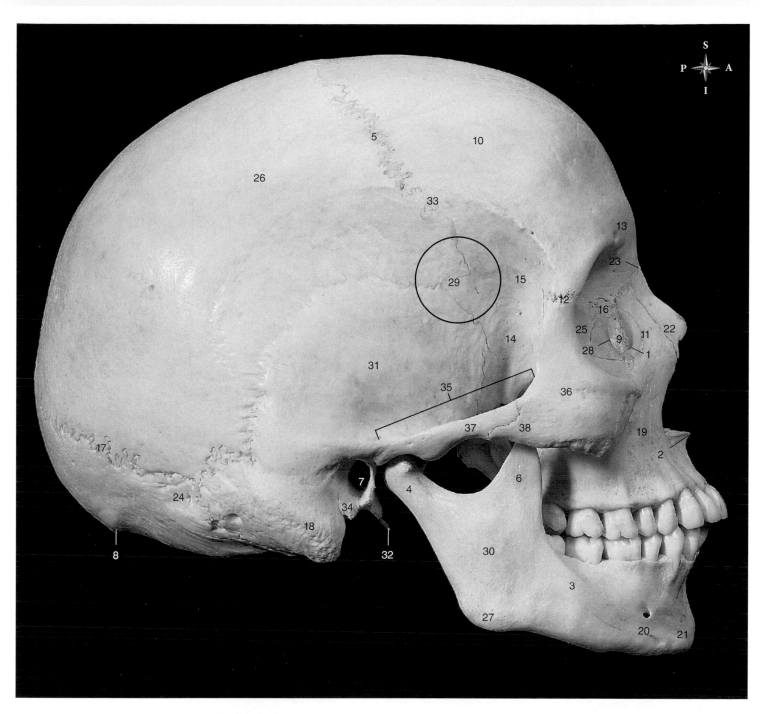

1 Anterior lacrimal crest
2 Anterior nasal spine
3 Body of mandible
4 Condyle of mandible
5 Coronal suture
6 Coronoid process of mandible
7 External acoustic meatus of temporal bone
8 External occipital protuberance (inion)
9 Fossa for lacrimal sac
10 Frontal bone
11 Frontal process of maxilla
12 Frontozygomatic suture
13 Glabella
14 Greater wing of sphenoid bone
15 Inferior temporal line
16 Lacrimal bone
17 Lambdoid suture
18 Mastoid process of temporal bone
19 Maxilla
20 Mental foramen
21 Mental protuberance
22 Nasal bone
23 Nasion
24 Occipital bone
25 Orbital part of ethmoid bone
26 Parietal bone
27 Angle of mandible
28 Posterior lacrimal crest
29 Pterion (encircled region)
30 Ramus of mandible
31 Squamous part of temporal bone
32 Styloid process of temporal bone
33 Superior temporal line
34 Tympanic part of temporal bone
35 Zygomatic arch
36 Zygomatic bone
37 Zygomatic process of temporal bone
38 Temporal process of zygomatic bone

Figure 2-4 *Skull (right lateral view).*

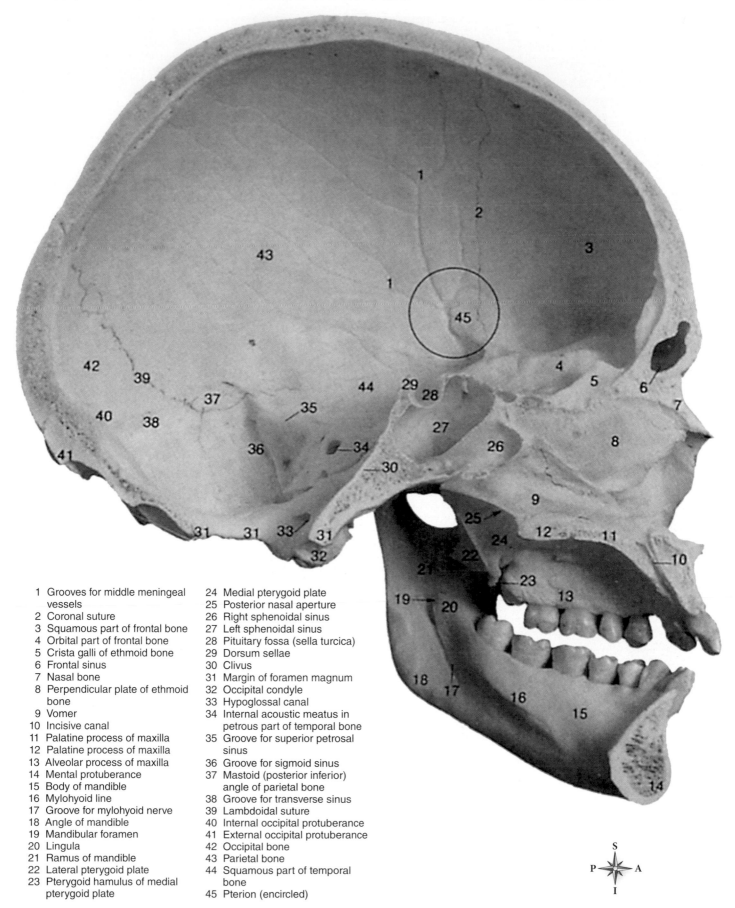

1 Grooves for middle meningeal
   vessels
2 Coronal suture
3 Squamous part of frontal bone
4 Orbital part of frontal bone
5 Crista galli of ethmoid bone
6 Frontal sinus
7 Nasal bone
8 Perpendicular plate of ethmoid
   bone
9 Vomer
10 Incisive canal
11 Palatine process of maxilla
12 Palatine process of maxilla
13 Alveolar process of maxilla
14 Mental protuberance
15 Body of mandible
16 Mylohyoid line
17 Groove for mylohyoid nerve
18 Angle of mandible
19 Mandibular foramen
20 Lingula
21 Ramus of mandible
22 Lateral pterygoid plate
23 Pterygoid hamulus of medial
   pterygoid plate

24 Medial pterygoid plate
25 Posterior nasal aperture
26 Right sphenoidal sinus
27 Left sphenoidal sinus
28 Pituitary fossa (sella turcica)
29 Dorsum sellae
30 Clivus
31 Margin of foramen magnum
32 Occipital condyle
33 Hypoglossal canal
34 Internal acoustic meatus in
   petrous part of temporal bone
35 Groove for superior petrosal
   sinus
36 Groove for sigmoid sinus
37 Mastoid (posterior inferior)
   angle of parietal bone
38 Groove for transverse sinus
39 Lambdoidal suture
40 Internal occipital protuberance
41 External occipital protuberance
42 Occipital bone
43 Parietal bone
44 Squamous part of temporal
   bone
45 Pterion (encircled)

Figure 2-5 *Left half of the skull (sagittal section).*

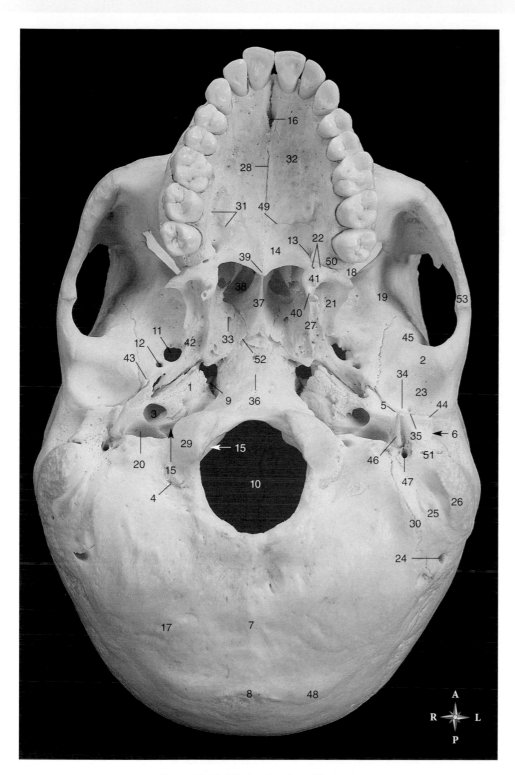

1 Apex of petrous part of temporal bone
2 Articular tubercle
3 Carotid canal
4 Condylar canal (posterior)
5 Edge of tegmen tympani
6 External acoustic meatus
7 External occipital crest
8 External occipital protuberance
9 Foramen lacerum
10 Foramen magnum
11 Foramen ovale
12 Foramen spinosum
13 Greater palatine foramen
14 Horizontal plate of palatine bone
15 Hypoglossal (anterior condylar) canal
16 Incisive fossa
17 Inferior nuchal line
18 Inferior orbital fissure (lit from below)
19 Infratemporal crest of greater wing of
   sphenoid bone
20 Jugular foramen
21 Lateral pterygoid plate
22 Lesser palatine foramina
23 Mandibular fossa
24 Mastoid foramen
25 Mastoid notch
26 Mastoid process
27 Medial pterygoid plate
28 Median palatine (intermaxillary) suture
29 Occipital condyle
30 Occipital groove
31 Palatine grooves and spines
32 Palatine process of maxilla
33 Palatinovaginal canal
34 Petrosquamous fissure
35 Petrotympanic fissure
36 Pharyngeal tubercle
37 Posterior border of vomer
38 Posterior nasal aperture (choana)
39 Posterior nasal spine
40 Pterygoid hamulus
41 Pyramidal process of palatine bone
42 Scaphoid fossa
43 Spine of sphenoid bone
44 Squamotympanic fissure
45 Squamous part of temporal bone
46 Styloid process
47 Stylomastoid foramen
48 Superior nuchal line
49 Transverse palatine (palatomaxillary) suture
50 Tuberosity of maxilla
51 Tympanic part of temporal bone
52 Vomerovaginal canal
53 Zygomatic arch

Figure 2-6 *Skull (external surface of the base).*

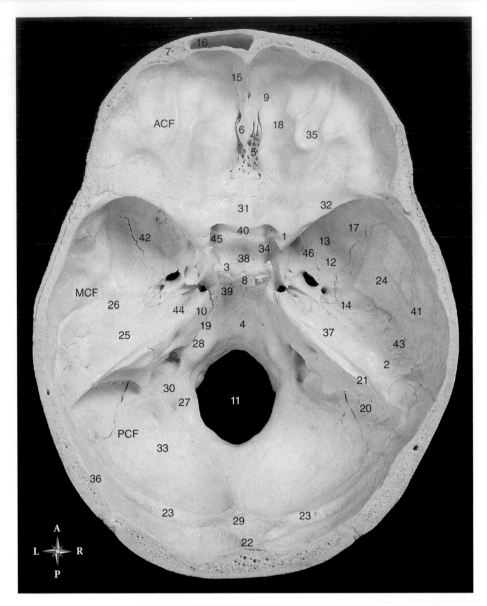

1 Anterior clinoid process
2 Arcuate eminence
3 Carotid groove
4 Clivus
5 Cribriform plate of
   ethmoid bone
6 Crista galli
7 Diploë
8 Dorsum sellae
9 Foramen cecum
10 Foramen lacerum
11 Foramen magnum
12 Foramen ovale
13 Foramen rotundum
14 Foramen spinosum
15 Frontal crest
16 Frontal sinus
17 Greater wing of
   sphenoid bone
18 Groove for anterior
   ethmoidal nerve and vessels
19 Groove for inferior
   petrosal sinus

20 Groove for sigmoid sinus
21 Groove for superior
   petrosal sinus
22 Groove for superior
   sagittal sinus
23 Groove for transverse
   sinus
24 Grooves for middle
   meningeal vessels
25 Hiatus and groove for
   greater petrosal nerve
26 Hiatus and groove for
   lesser petrosal nerve
27 Hypoglossal canal
28 Internal acoustic meatus
29 Internal occipital
   protuberance
30 Jugular foramen
31 Jugum of sphenoid bone
32 Lesser wing of
   sphenoid bone
33 Occipital bone
34 Optic canal

35 Orbital part of frontal bone
36 Parietal bone
   (posteroinferior angle only)
37 Petrous part of temporal
   bone
38 Pituitary fossa
   (sella turcica)
39 Posterior clinoid process
40 Prechiasmatic groove
41 Squamous part of
   temporal bone
42 Superior orbital fissure
43 Tegmen tympani
44 Trigeminal impression
45 Tuberculum sellae
46 Venous foramen

ACF, Anterior cranial fossa
MCF, Middle cranial fossa
PCF, Posterior cranial fossa

Figure 2-7 *Skull (internal surface of the base).*

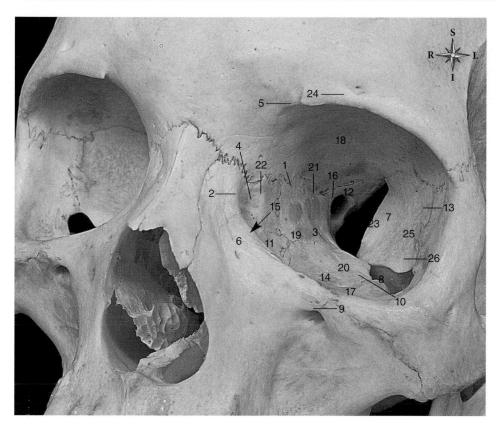

1 Anterior ethmoidal foramen
2 Anterior lacrimal crest
3 Body of sphenoid bone, forming medial wall
4 Fossa for lacrimal sac
5 Frontal notch
6 Frontal process of maxilla, forming medial wall
7 Greater wing of sphenoid bone, forming lateral wall
8 Inferior orbital fissure
9 Infraorbital foramen
10 Infraorbital groove
11 Lacrimal bone, forming medial wall
12 Lesser wing of sphenoid bone, forming roof
13 Marginal tubercle
14 Maxilla, forming floor
15 Nasolacrimal canal
16 Optic canal
17 Orbital border of zygomatic bone, forming floor
18 Orbital part of frontal bone, forming roof
19 Orbital plate of ethmoid bone, forming medial wall
20 Orbital process of palatine bone, forming floor
21 Posterior ethmoidal foramen
22 Posterior lacrimal crest
23 Superior orbital fissure
24 Supraorbital foramen
25 Zygomatic bone, forming lateral wall
26 Zygomatico-orbital foramen

Figure 2-8 *Skull (bones of the eye orbit).*

1 Air cells of ethmoidal sinus
2 Clivus
3 Cribriform plate of ethmoid bone
4 Dorsum sellae
5 Ethmoidal bulla
6 Frontal sinus
7 Horizontal plate of palatine bone
8 Incisive canal
9 Inferior meatus
10 Inferior nasal concha
11 Lateral pterygoid plate
12 Left sphenoidal sinus
13 Medial pterygoid plate
14 Nasal bone
15 Nasal spine of frontal bone
16 Opening of maxillary sinus
17 Palatine process of maxilla
18 Perpendicular plate of palatine bone
19 Pituitary fossa (sella turcica)
20 Pterygoid hamulus
21 Right sphenoidal sinus
22 Semilunar hiatus
23 Sphenopalatine foramen
24 Uncinate process of ethmoid bone

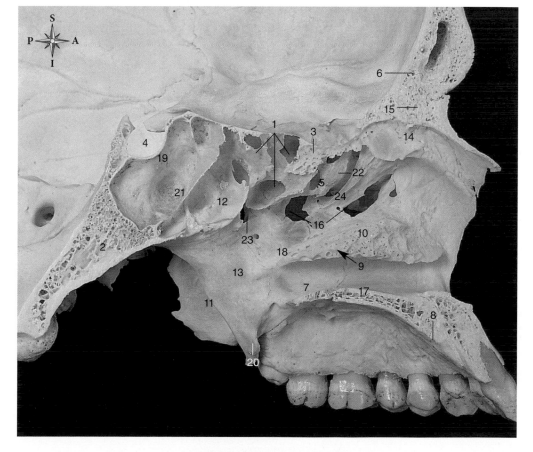

Figure 2-9 *Nasal cavity (lateral wall).*

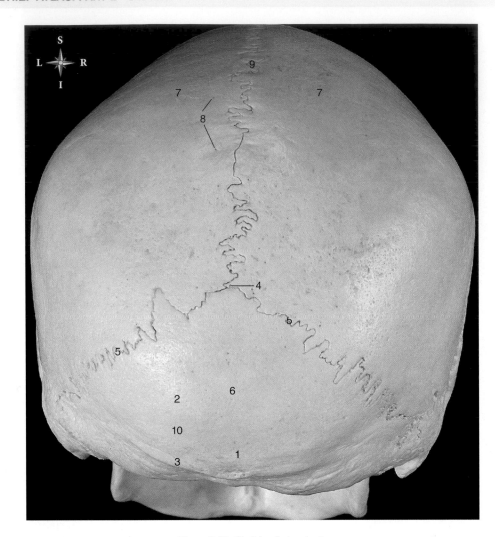

1 External occipital protuberance (inion)
2 Highest nuchal line
3 Inferior nuchal line
4 Lambda
5 Lambdoid suture
6 Occipital bone
7 Parietal bone
8 Parietal foramen
9 Sagittal suture
10 Superior nuchal line

Figure 2-10 *Skull (posterior view).*

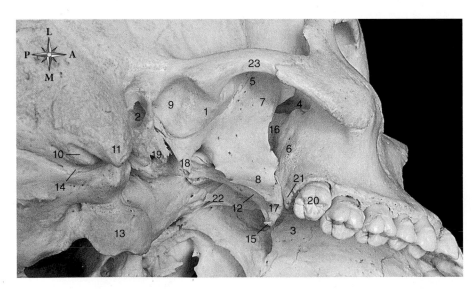

1 Articular tubercle
2 External acoustic meatus
3 Horizontal plate of palatine bone
4 Inferior orbital fissure
5 Infratemporal crest
6 Infratemporal (posterior) surface of maxilla
7 Infratemporal surface of greater wing of sphenoid bone
8 Lateral pterygoid plate
9 Mandibular fossa
10 Mastoid notch
11 Mastoid process
12 Medial pterygoid plate
13 Occipital condyle
14 Occipital groove
15 Pterygoid hamulus
16 Pterygomaxillary fissure and pterygopalatine fossa
17 Pyramidal process of palatine bone
18 Spine of sphenoid bone
19 Styloid process and sheath
20 Third molar tooth
21 Tuberosity of maxilla
22 Vomer
23 Zygomatic arch

Figure 2-11 *Skull (oblique inferior view).*

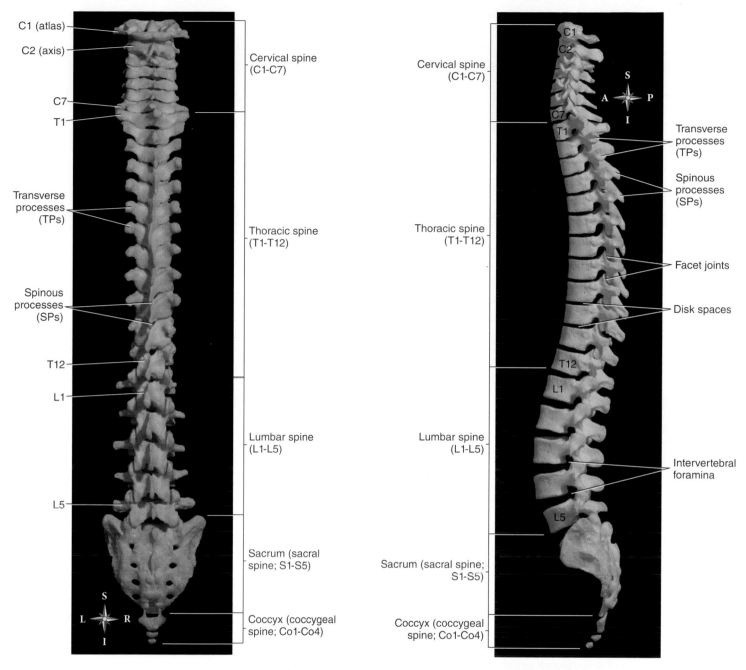

C1 (atlas)

C2 (axis)

C7

T1

Transverse
processes
(TPs)

Spinous
processes
(SPs)

T12

L1

L5

Cervical spine
(C1-C7)

Thoracic spine
(T1-T12)

Lumbar spine
(L1-L5)

Sacrum (sacral
spine; S1-S5)

Coccyx (coccygeal
spine; Co1-Co4)

Figure 2-12 *Vertebral column (posterior view).*

Cervical spine
(C1-C7)

Thoracic spine
(T1-T12)

Lumbar spine
(L1-L5)

Sacrum (sacral spine;
S1-S5)

Coccyx (coccygeal
spine; Co1-Co4)

C1
C2

C7
T1

T12
L1

L5

Transverse
processes
(TPs)

Spinous
processes
(SPs)

Facet joints

Disk spaces

Intervertebral
foramina

Figure 2-13 *Vertebral column (right lateral view).*

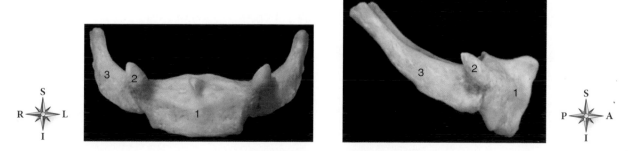

1 Body 2 Lesser cornu 3 Greater cornu

Figure 2-14 *Hyoid bone (anterior view [left]; right lateral view [right]).*

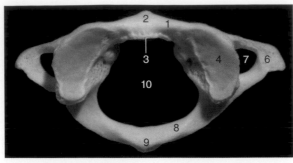

1 Anterior arch
2 Anterior tubercle
3 Facet for dens of axis (C2)
4 Superior articular process/facet
5 Inferior articular process/facet
6 Transverse process (TP)
7 Transverse foramen
8 Posterior arch
9 Posterior tubercle
10 Vertebral foramen
11 Lateral mass

Figure 2-15 *First cervical vertebra (superior view).*

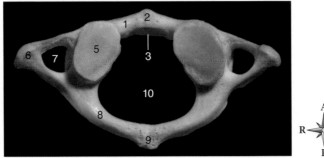

Figure 2-16 *First cervical vertebra (inferior view).*

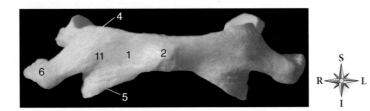

Figure 2-17 *First cervical vertebra (anterior view).*

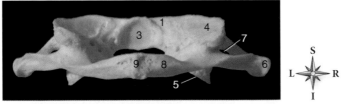

Figure 2-18 *First cervical vertebra (posterior view).*

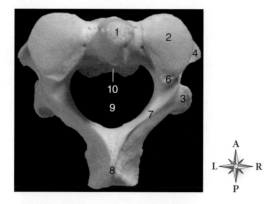

Figure 2-19 *Second cervical vertebra (C2; axis; superior view).*

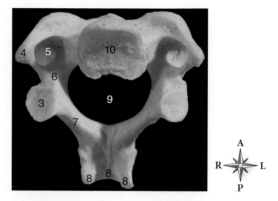

Figure 2-20 *Second cervical vertebra (C2; axis; inferior view).*

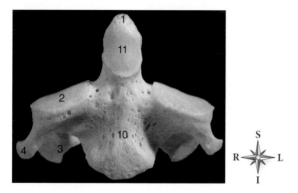

Figure 2-21 *Second cervical vertebra (C2; axis; anterior view).*

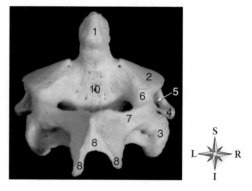

Figure 2-22 *Second cervical vertebra (C2; axis; posterior view).*

1 Dens (odontoid process)
2 Superior articular process/facet
3 Inferior articular process/facet
4 Transverse process (TP)
5 Transverse foramen
6 Pedicle
7 Lamina
8 Spinous process (SP) (bifid)
9 Vertebral foramen
10 Body
11 Facet on dens

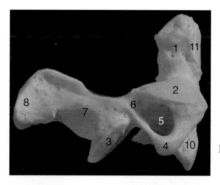

Figure 2-23 *Second cervical vertebra (C2; axis; right lateral view).*

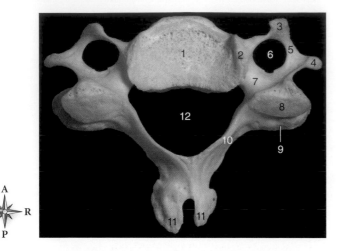

Figure 2-24 *Typical cervical vertebra (C5; superior view).*

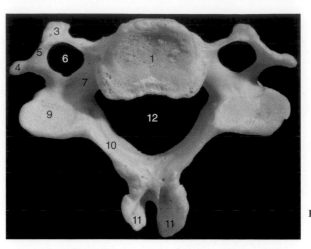

Figure 2-25 *Typical cervical vertebra (C5; inferior view).*

Figure 2-26 *Typical cervical vertebra (C5; anterior view).*

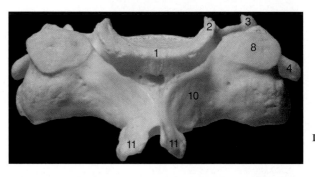

Figure 2-27 *Typical cervical vertebra (C5; posterior view).*

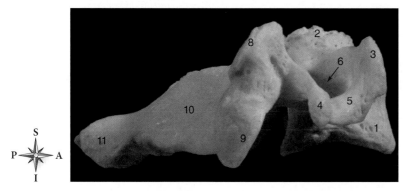

Figure 2-28 *Typical cervical vertebra (C5; right lateral view).*

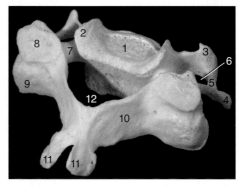

Figure 2-29 *Typical cervical vertebra (C5; oblique view [from above right]).*

1 Body
2 Uncus of body
3 Anterior tubercle of transverse process (TP)
4 Posterior tubercle of TP
5 Groove for spinal nerve (on TP)
6 Transverse foramen
7 Pedicle
8 Superior articular process/facet
9 Inferior articular process/facet
10 Lamina
11 Spinous process (SP) (bifid)
12 Vertebral foramen

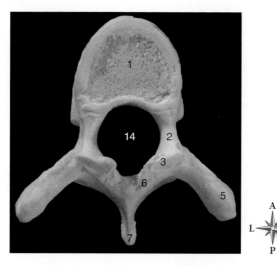

Figure 2-30 *Typical thoracic vertebra (T5; superior view).*

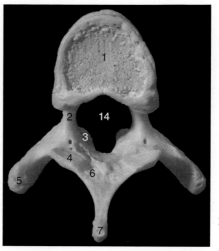

Figure 2-31 *Typical thoracic vertebra (T5; inferior view).*

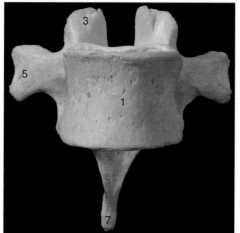

Figure 2-32 *Typical thoracic vertebra (T5; anterior view).*

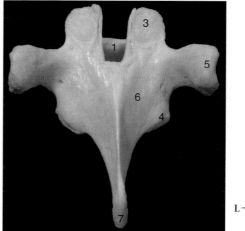

Figure 2-33 *Typical thoracic vertebra (T5; posterior view).*

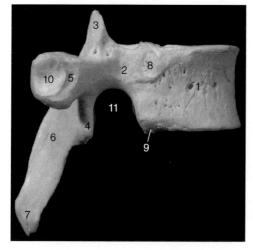

Figure 2-34 *Typical thoracic vertebra (T5; right lateral view).*

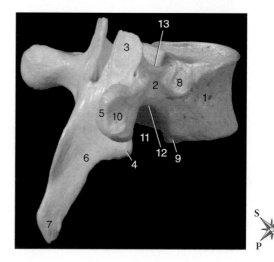

Figure 2-35 *Typical thoracic vertebra (T5; oblique view [from posterior right]).*

1 Body	8 Superior costal hemifacet
2 Pedicle	9 Inferior costal hemifacet
3 Superior articular process/facet	10 Transverse costal facet
4 Inferior articular process/facet	11 Intervertebral foramen
5 Transverse process (TP)	12 Inferior vertebral notch
6 Lamina	13 Superior vertebral notch
7 Spinous process (SP)	14 Vertebral foramen

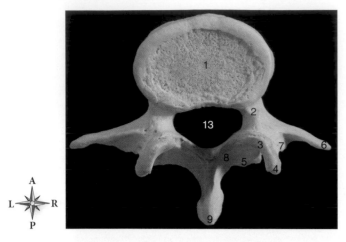

Figure 2-36 *Typical lumbar vertebra (L3; superior view).*

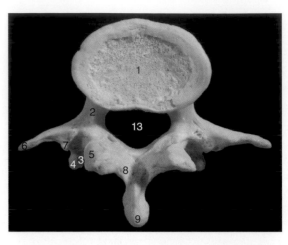

Figure 2-37 *Typical lumbar vertebra (L3; inferior view).*

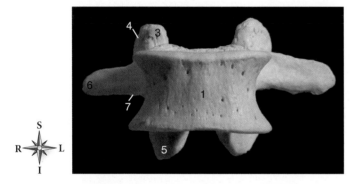

Figure 2-38 *Typical lumbar vertebra (L3; anterior view).*

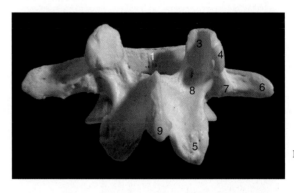

Figure 2-39 *Typical lumbar vertebra (L3; posterior view).*

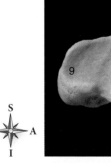

Figure 2-40 *Typical lumbar vertebra (L3; right lateral view).*

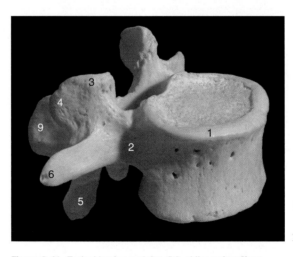

Figure 2-41 *Typical lumbar vertebra (L3; oblique view [from anterior right]).*

1 Body
2 Pedicle
3 Superior articular process/facet
4 Mamillary process
5 Inferior articular process/facet
6 Transverse process (TP)
7 Accessory process
8 Lamina
9 Spinous process (SP)
10 Intervertebral foramen
11 Inferior vertebral notch
12 Superior vertebral notch
13 Vertebral foramen

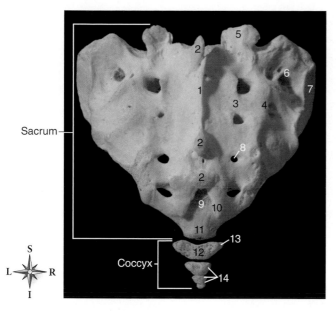

Figure 2-42 *Sacrum and coccyx (posterior view).*

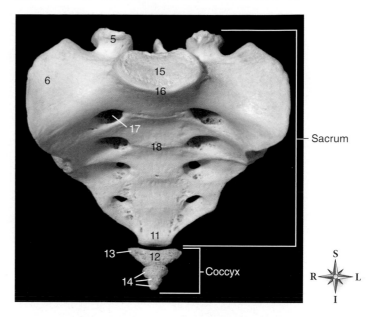

Figure 2-43 *Sacrum and coccyx (anterior view).*

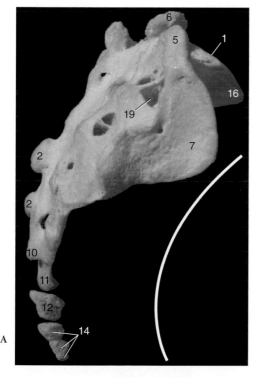

Figure 2-44 *Sacrum and coccyx (right lateral view).*

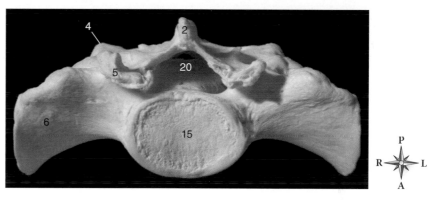

Figure 2-45 *Sacrum and coccyx (superior view).*

1 Median sacral crest
2 Tubercles along the median sacral crest
3 Intermediate sacral crest
4 Lateral sacral crest
5 Superior articular process/facet
6 Ala (wing)
7 Auricular surface (articular surface for ilium)
8 3rd Posterior foramen
9 Sacral hiatus
10 Sacral cornu
11 Apex
12 1st Coccygeal element
13 Coccygeal transverse process (TP)
14 2nd to 4th Coccygeal elements (fused)
15 Sacral base
16 Sacral promontory
17 1st Anterior foramen
18 Fusion of 2nd and 3rd sacral vertebrae
19 First posterior foramen
20 Sacral canal

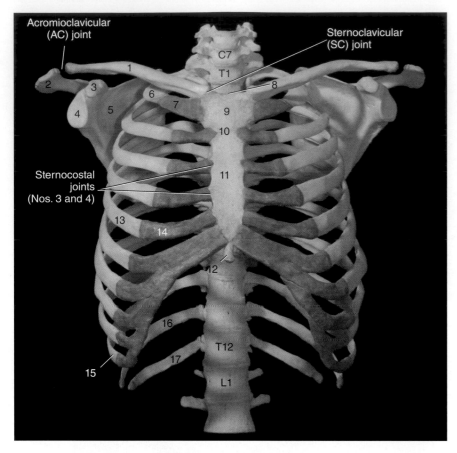

Figure 2-46 *Ribcage (anterior view).*

1 Clavicle
2 Acromion
3 Coracoid process
4 Glenoid fossa
5 Subscapular fossa
6 1st rib
7 Cartilage of 1st rib
8 Sternal notch
9 Manubrium of sternum
10 Sternal angle
11 Body of sternum
12 Xiphoid process of sternum
13 5th rib
14 Cartilage of 5th rib
15 10th rib
16 11th rib
17 12th rib
18 Clavicular notch of the manubrium
19 Notch for 1st costal cartilage
20 Notch for 2nd costal cartilage
21 Notch for 3rd costal cartilage
22 Notch for 4th costal cartilage
23 Notch for 5th costal cartilage
24 Notch for 6th costal cartilage
25 Notch for 7th costal cartilage

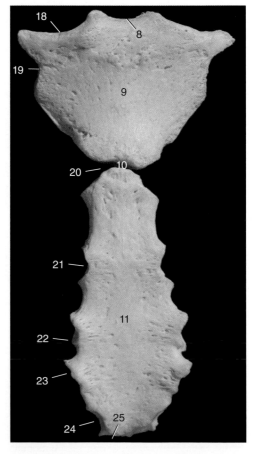

Figure 2-47 *Sterum.*

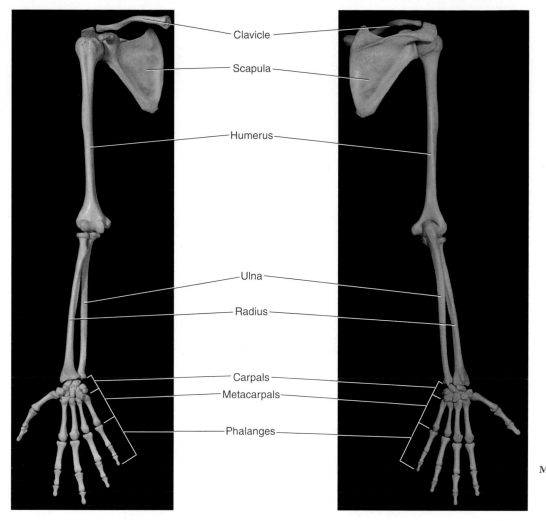

Figure 2-48 *Upper extremity (anterior view).*

Figure 2-49 *Upper extremity (posterior view).*

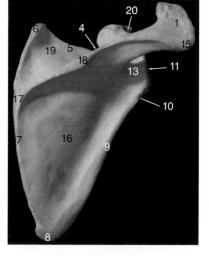

Figure 2-50 *Right scapula (anterior view).*

Figure 2-51 *Right scapula (posterior view).*

1 Acromion	8 Inferior angle	15 Acromial organizer
2 Apex of coracoid process	9 Lateral border	16 Infraspinous fossa
3 Base of coracoid process	10 Infraglenoid tubercle	17 Root of the spine
4 Suprascapular notch	11 Glenoid fossa	18 Spine
5 Superior border	12 Supraglenoid tubercle	19 Supraspinous fossa
6 Superior angle	13 Neck	20 Coracoid process
7 Medial border	14 Subscapular fossa	

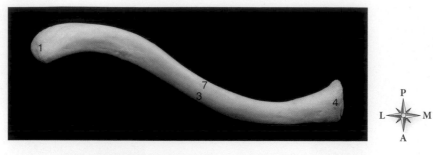

Figure 2-52 *Right clavicle (superior view).*

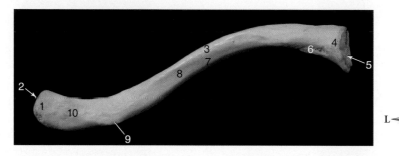

Figure 2-53 *Right clavicle (inferior view).*

Figure 2-54 *Right clavicle (anterior view).*

Figure 2-55 *Right clavicle (posterior view).*

1 Acromial end
2 Articular surface for acromioclavicular (AC) joint
3 Anterior border
4 Sternal end
5 Articular surface for sternoclavicular joint
6 Costal tubercle
7 Posterior border
8 Subclavian groove
9 Conoid tubercle
10 Trapezoid line
11 Superior border
12 Inferior border

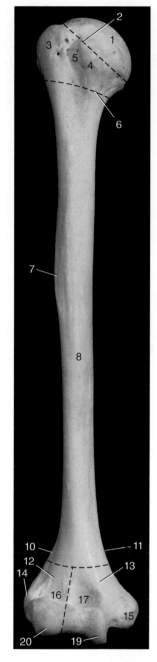

Figure 2-56 *Right humerus (anterior view).*

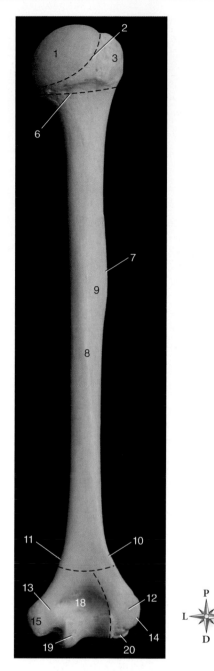

Figure 2-57 *Right humerus (posterior view).*

1	Head	11	Medial supracondylar ridge
2	Anatomic neck	12	Lateral condyle
3	Greater tubercle	13	Medial condyle
4	Lesser tubercle	14	Lateral epicondyle
5	Bicipital groove	15	Medial epicondyle
6	Surgical neck	16	Radial fossa
7	Deltoid tuberosity	17	Coronoid fossa
8	Body (shaft)	18	Olecranon fossa
9	Groove for radial nerve	19	Trochlea
10	Lateral supracondylar ridge	20	Capitulum

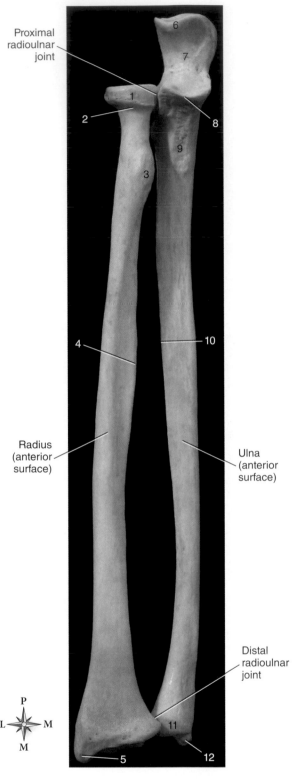

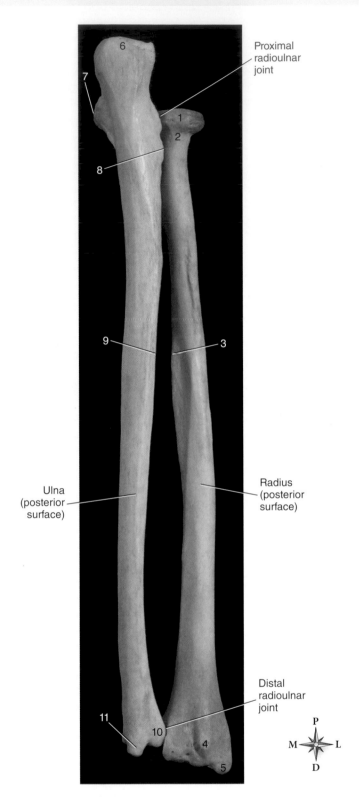

Figure 2-58 *Right radius and ulna (anterior view).*

Figure 2-59 *Right radius and ulna (posterior view).*

Landmarks of the Radius
1 Head
2 Neck
3 Tuberosity
4 Interosseous crest
5 Styloid process

Landmarks of the Ulna
6 Olecranon process
7 Trochlear notch
8 Coronoid process
9 Tuberosity
10 Interosseous crest
11 Head
12 Styloid process

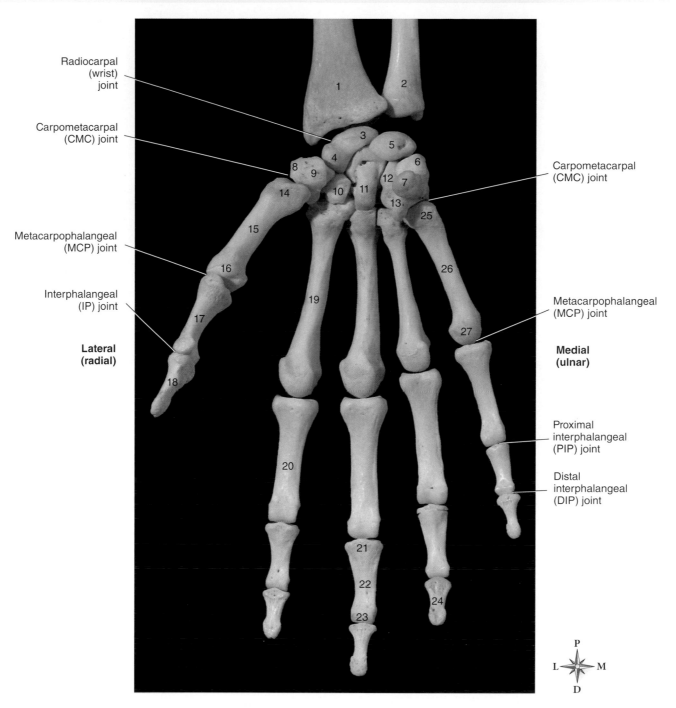

Radiocarpal (wrist) joint

Carpometacarpal (CMC) joint

Metacarpophalangeal (MCP) joint

Interphalangeal (IP) joint

Lateral (radial)

Carpometacarpal (CMC) joint

Metacarpophalangeal (MCP) joint

Medial (ulnar)

Proximal interphalangeal (PIP) joint

Distal interphalangeal (DIP) joint

Figure 2-60 *Wrist and hand (anterior view).*

1 Radius
2 Ulna
3 Scaphoid
4 Tubercle of scaphoid
5 Lunate
6 Triquetrum
7 Pisiform
8 Trapezium
9 Tubercle of trapezium
10 Trapezoid
11 Capitate
12 Hamate
13 Hook of hamate
14 Base of 1st metacarpal (of thumb)

15 Body (shaft) of 1st metacarpal (of thumb)
16 Head of 1st metacarpal (of thumb)
17 Proximal phalanx of thumb
18 Distal phalanx of thumb
19 2nd Metacarpal (of index finger)
20 Proximal phalanx of index finger
21 Base of middle phalanx of middle finger
22 Body (shaft) of middle phalanx of middle finger
23 Head of middle phalanx of middle finger
24 Distal phalanx of ring finger
25 Base of 5th metacarpal (of little finger)
26 Body (shaft) of 5th metacarpal (of little finger)
27 Head of 5th metacarpal (of little finger)

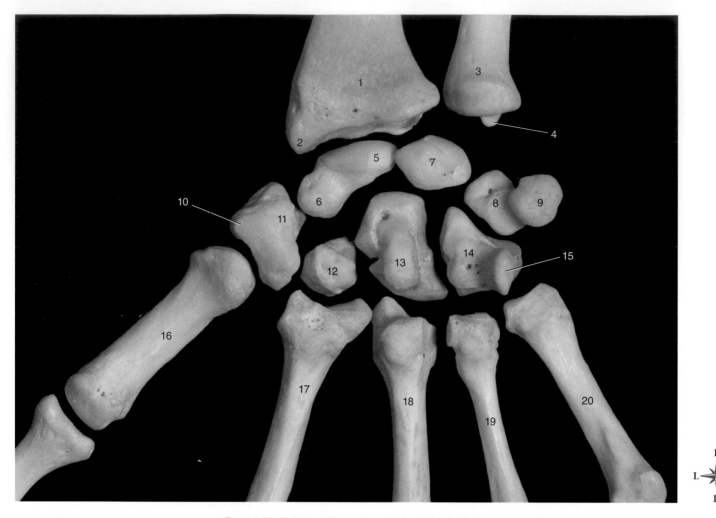

Figure 2-61 *Right carpal bones (separated; anterior view).*

1 Radius	11 Tubercle of trapezium
2 Styloid process of radius	12 Trapezoid
3 Ulna	13 Capitate
4 Styloid process of ulna	14 Hamate
5 Scaphoid	15 Hook of hamate
6 Tubercle of scaphoid	16 1st Metacarpal (of thumb)
7 Lunate	17 2nd Metacarpal (of index finger)
8 Triquetrum	18 3rd Metacarpal (of middle finger)
9 Pisiform	19 4th Metacarpal (of ring finger)
10 Trapezium	20 5th Metacarpal (of little finger)

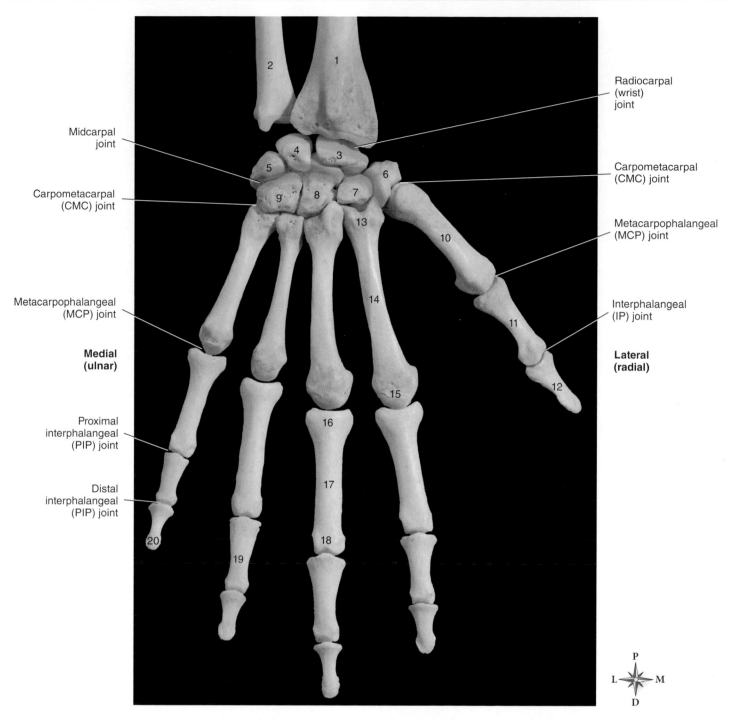

Radiocarpal
(wrist)
joint

Midcarpal
joint

Carpometacarpal
(CMC) joint

Carpometacarpal
(CMC) joint

Metacarpophalangeal
(MCP) joint

Metacarpophalangeal
(MCP) joint

Interphalangeal
(IP) joint

**Medial
(ulnar)**

**Lateral
(radial)**

Proximal
interphalangeal
(PIP) joint

Distal
interphalangeal
(PIP) joint

Figure 2-62 *Right wrist and hand (posterior view).*

1 Radius	11 Proximal phalanx of thumb
2 Ulna	12 Distal phalanx of thumb
3 Scaphoid	13 Base of metacarpal of index finger
4 Lunate	14 Body (shaft) of metacarpal of index finger
5 Triquetrum	15 Head of metacarpal of index finger
6 Trapezium	16 Base of proximal phalanx of middle finger
7 Trapezoid	17 Body (shaft) of proximal phalanx of middle finger
8 Capitate	18 Head of proximal phalanx of middle finger
9 Hamate	19 Middle phalanx of little finger
10 1st Metacarpal (of thumb)	20 Distal phalanx of little finger

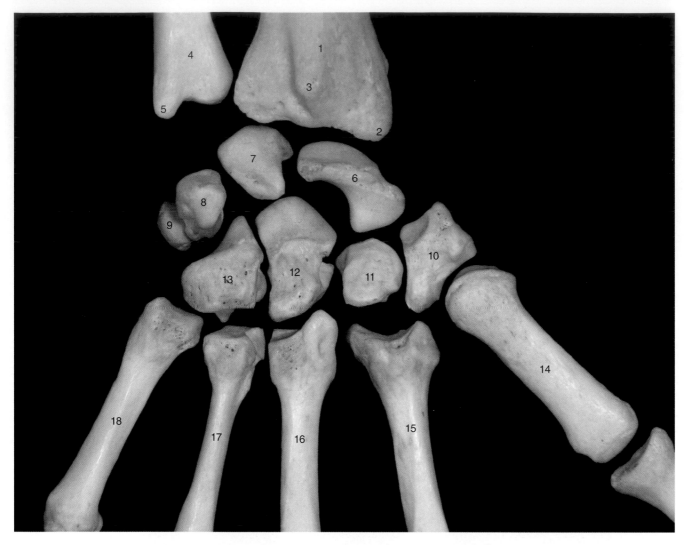

Figure 2-63 *Right carpal bones (separated; posterior view).*

1 Radius	10 Trapezium
2 Styloid process of radius	11 Trapezoid
3 Dorsal tubercle of radius	12 Capitate
4 Ulna	13 Hamate
5 Styloid process of ulna	14 1st Metacarpal (of thumb)
6 Scaphoid	15 2nd Metacarpal (of index finger)
7 Lunate	16 3rd Metacarpal (of middle finger
8 Triquetrum	17 4th Metacarpal (of ring finger)
9 Pisiform	18 5th Metacarpal (of little finger)

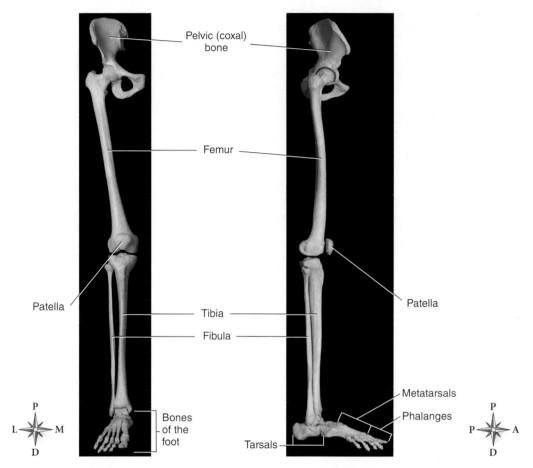

Figure 2-64 *Right lower extremity (anterior view).*

Figure 2-65 *Right lower extremity (posterior view).*

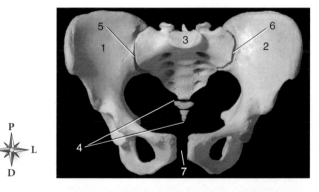

Figure 2-66 *Pelvis (anterior view).*

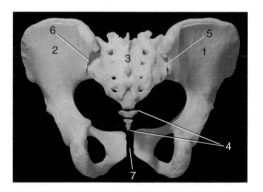

Figure 2-67 *Pelvis (posterior view).*

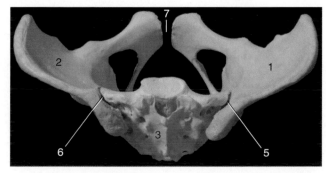

Figure 2-68 *Pelvis (superior view).*

1 Right pelvic (coxal) bone
2 Left pelvic (coxal) bone
3 Sacrum
4 Coccyx
5 Right sacroiliac (SI) joint
6 Left SI joint
7 Pubic symphysis joint

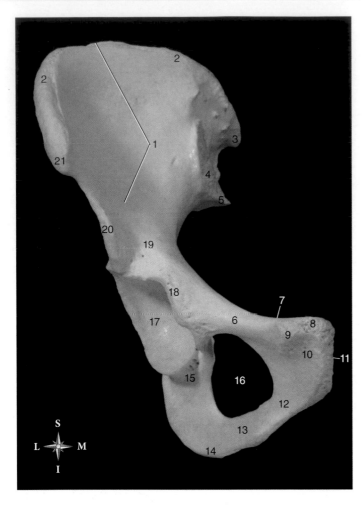

1 Wing of the ilium (iliac fossa on internal surface)
2 Iliac crest
3 Posterior superior iliac spine (PSIS)
4 Articular surface for sacroiliac joint
5 Posterior inferior iliac spine (PIIS)
6 Superior ramus of pubis
7 Pectineal line of pubis
8 Pubic crest
9 Pubic tubercle
10 Body of pubis
11 Articular surface for pubic symphysis
12 Inferior ramus of pubis
13 Ramus of ischium
14 Ischial tuberosity
15 Body of ischium
16 Obturator foramen
17 Acetabulum
18 Rim of acetabulum
19 Body of ilium
20 Anterior inferior iliac spine (AIIS)
21 Anterior superior iliac spine (ASIS)

Figure 2-69 *Right pelvic bone (coxal; anterior view).*

1 Wing of the ilium (iliac fossa on internal surface)
2 Iliac crest
3 Anterior superior iliac spine (ASIS)
4 Rim of acetabulum
5 Ischial spine
6 Body of ischium
7 Ischial tuberosity
8 Ramus of ischium
9 Inferior ramus of pubis
10 Body of pubis
11 Superior ramus of pubis
12 Pectineal line of pubis
13 Obturator foramen
14 Body of ilium
15 Posterior inferior iliac spine (PIIS)
16 Posterior superior iliac spine (PSIS)
17 Inferior gluteal line (*dashed line*)
18 Anterior gluteal line (*dashed line*)
19 Posterior gluteal line (*dashed line*)
20 Greater sciatic notch
21 Lesser sciatic notch

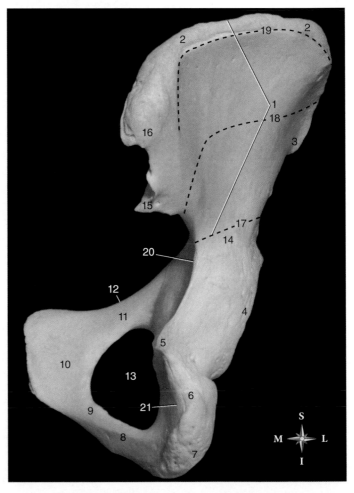

Figure 2-70 *Right pelvic bone (coxal; posterior view).*

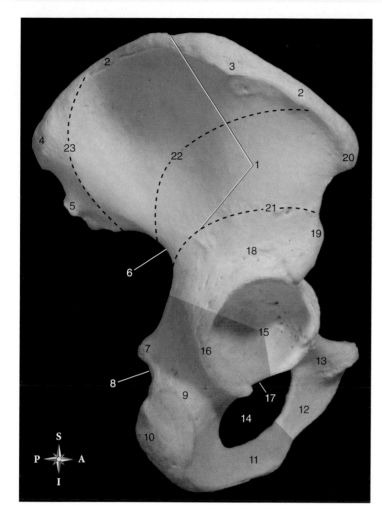

1 Wing of the ilium (external/gluteal surface)
2 Iliac crest
3 Tubercle of the iliac crest
4 Posterior superior iliac spine (PSIS)
5 Posterior inferior iliac spine (PIIS)
6 Greater sciatic notch
7 Ischial spine
8 Lesser sciatic notch
9 Body of ischium
10 Ischial tuberosity
11 Ramus of ischium
12 Inferior ramus of pubis
13 Body of pubis
14 Obturator foramen
15 Acetabulum
16 Rim of acetabulum
17 Notch of acetabulum
18 Body of ilium
19 Anterior inferior iliac spine (AIIS)
20 Anterior superior iliac spine (ASIS)
21 Inferior gluteal line *(dashed line)*
22 Anterior gluteal line *(dashed line)*
23 Posterior gluteal line *(dashed line)*

Figure 2-71 *Right pelvic bone (coxal; lateral view).* Blue, Ilium; pink, ischium; yellow, pubis.

1 Wing of the ilium (iliac fossa on internal surface)
2 Iliac crest
3 Posterior superior iliac spine (PSIS)
4 Iliac tuberosity
5 Articular surface of ilium for sacroiliac joint
6 Posterior inferior iliac spine (PIIS)
7 Greater sciatic notch
8 Ischial spine
9 Lesser sciatic notch
10 Body of ischium
11 Ischial tuberosity
12 Ramus of ischium
13 Inferior ramus of pubis
14 Articular surface of pubis for pubis symphysis
15 Pubic tubercle
16 Superior ramus of pubis
17 Pectineal line of pubis
18 Body of pubis
19 Iliopectineal line
20 Arcuate line of the ilium
21 Obturator foramen
22 Body of ilium
23 Anterior inferior iliac spine (AIIS)
24 Anterior superior iliac spine (ASIS)

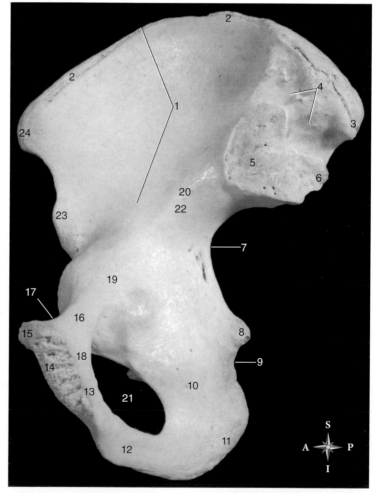

Figure 2-72 *Right pelvic bone (coxal; medial view).*

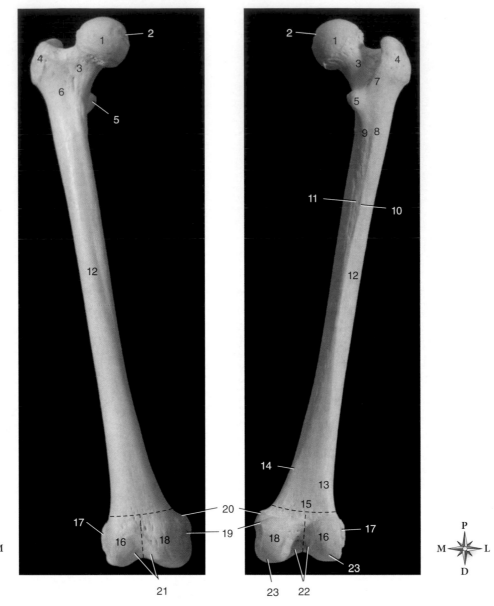

Figure 2-73 *Right femur (anterior view).*   **Figure 2-74** *Right femur (posterior view).*

 1 Head
 2 Fovea of the head
 3 Neck
 4 Greater trochanter
 5 Lesser trochanter
 6 Intertrochanteric line
 7 Intertrochanteric crest
 8 Gluteal tuberosity
 9 Pectineal line
10 Lateral lip of linea aspera
11 Medial lip of linea aspera
12 Body (shaft)

13 Lateral supracondylar line
14 Medial supracondylar line
15 Popliteal surface
16 Lateral condyle
17 Lateral epicondyle
18 Medial condyle
19 Medial epicondyle
20 Adductor tubercle
21 Articular surface for patellofemoral joint
22 Intercondylar fossa
23 Articular surface for knee (tibiofemoral) joint

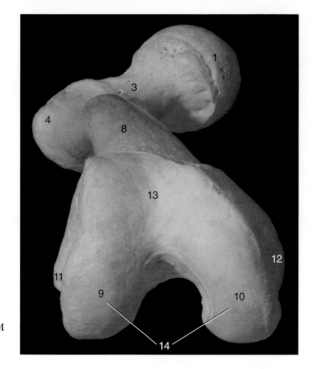

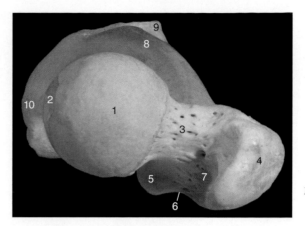

Figure 2-76 *Femur (proximal [superior] view).*

1 Head
2 Fovea of the head
3 Neck
4 Greater trochanter
5 Lesser trochanter
6 Intertrochanteric crest
7 Trochanteric fossa
8 Body (shaft), anterior surface
9 Lateral condyle
10 Medial condyle
11 Lateral epicondyle
12 Medial epicondyle
13 Articular surface for patellofemoral joint
14 Articular surface for knee (tibiofemoral) joint

Figure 2-75 *Femur (distal [inferior] view).*

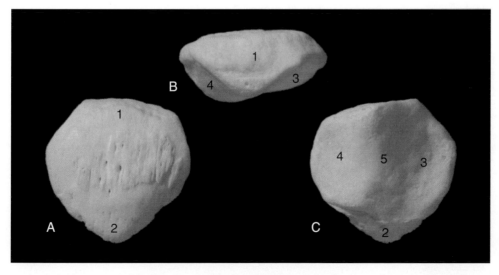

Figure 2-77 *Right patella (A, anterior view; B, proximal [superior] view; C, posterior view).*

1 Base
2 Apex
3 Facet for lateral condyle of femur
4 Facet for medial condyle of femur
5 Vertical ridge

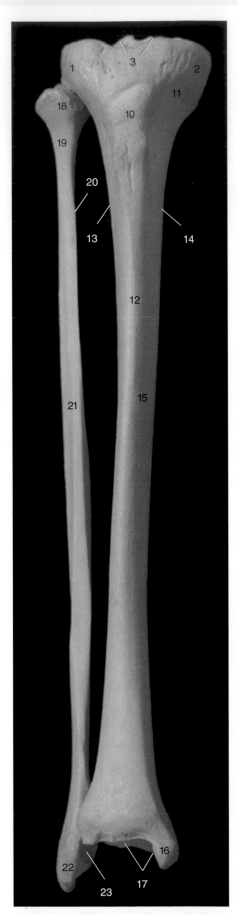

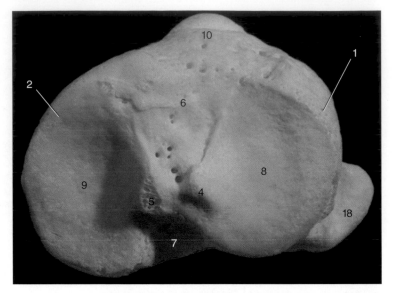

Figure 2-79 *Right tibia and fibula (proximal view).*

Tibial Landmarks:
1 Lateral condyle
2 Medial condyle
3 Intercondylar eminence
4 Lateral tubercle of intercondylar eminence
5 Medial tubercle of intercondylar eminence
6 Anterior intercondylar area
7 Posterior intercondylar area
8 Lateral facet (articular surface for knee [i.e., tibiofemoral] joint)
9 Medial facet (articular surface for knee [i.e., tibiofemoral] joint)
10 Tuberosity
11 Impression for iliotibial tract
12 Crest (i.e., anterior border)
13 Interosseous border
14 Medial border
15 Body (shaft)
16 Medial malleolus
17 Articular surfaces for ankle joint

Fibular Landmarks:
18 Head
19 Neck
20 Interosseous border
21 Body (shaft)
22 Lateral malleolus
23 Articular surface for ankle joint

Figure 2-78 *Right tibia and fibula (anterior view).*

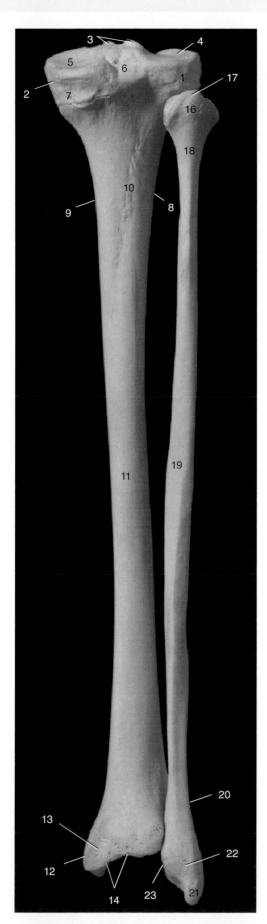

Figure 2-80 *Right tibia and fibula (posterior view).*

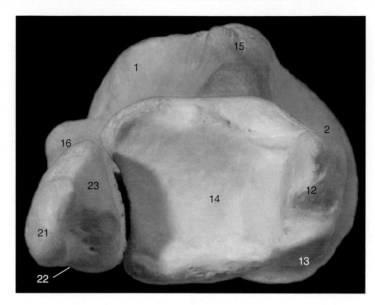

Figure 2-81 *Right tibia and fibula (distal view).*

Tibial Landmarks:
1 Lateral condyle
2 Medial condyle
3 Intercondylar eminence
4 Lateral facet (articular surface for knee [i.e., tibiofemoral] joint)
5 Medial facet (articular surface for knee [i.e., tibiofemoral] joint)
6 Posterior intercondylar area
7 Groove for semimembranosus muscle
8 Interosseous border
9 Medial border
10 Soleal line
11 Body (shaft)
12 Medial malleolus
13 Groove for tibialis posterior
14 Articular surfaces for ankle joint
15 Tuberosity

Fibular Landmarks:
16 Head
17 Apex of head
18 Neck
19 Body (shaft)
20 Lateral surface
21 Lateral malleolus
22 Groove for fibularis brevis
23 Articular surface for ankle joint

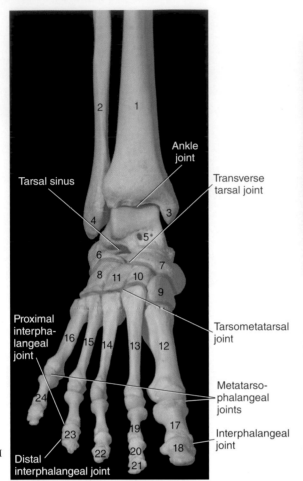

Figure 2-82 *Right ankle (anterior view).*

Figure 2-83 *Right ankle (lateral view).*

1 Tibia	13 2nd Metatarsal
2 Fibula	14 3rd Metatarsal
3 Medial malleolus (of tibia)	15 4th Metatarsal
4 Lateral malleolus (of fibula)	16 5th Metatarsal
5 Talus	17 Proximal phalanx of big toe
6 Calcaneus	18 Distal phalanx of big toe
7 Navicular	19 Proximal phalanx of 2nd toe
8 Cuboid	20 Middle phalanx of 2nd toe
9 1st Cuneiform	21 Distal phalanx of 2nd toe
10 2nd Cuneiform	22 Distal phalanx of 3rd toe
11 3rd Cuneiform	23 Middle phalanx of 4th toe
12 1st Metatarsal	24 Proximal phalanx of little toe (i.e., 5th toe)

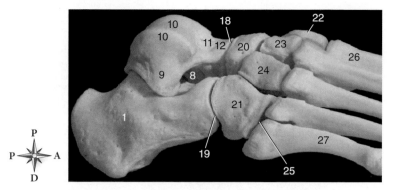

Figure 2-84 *Right ankle (subtalar joint; right lateral view).*

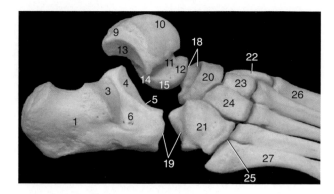

Figure 2-85 *Right ankle (subtalar joint; right lateral view; subtalar joint open).*

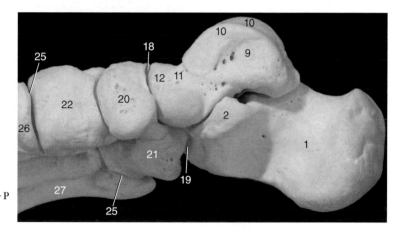

Figure 2-86 *Right ankle (subtalar joint; medial view).*

1 Calcaneus (# I-7)
2 Sustentaculum tali
3 Calcaneal posterior facet (of subtalar joint)
4 Calcaneal middle facet (of subtalar joint)
5 Calcaneal anterior facet (of subtalar joint)
6 Sulcus (of calcaneus)
7 Articular surface for calcaneocuboid joint (of transverse tarsal joint)
8 Tarsal sinus
9 Talus (#9-17)
10 Articular surface for ankle joint
11 Neck of talus
12 Head of talus
13 Talar posterior facet (of subtalar joint)
14 Talar middle facet (of subtalar joint)
15 Talar anterior facet (of subtalar joint)
16 Sulcus (of talus)
17 Articular surface for talonavicular joint (of transverse tarsal joint)
18 Talonavicular joint (of transverse tarsal joint)
19 Calcaneocuboid joint (of transverse tarsal joint)
20 Navicular
21 Cuboid
22 1st Cuneiform
23 2nd Cuneiform
24 3rd Cuneiform
25 Tarsometatarsal joint
26 1st Metatarsal
27 5th Metatarsal

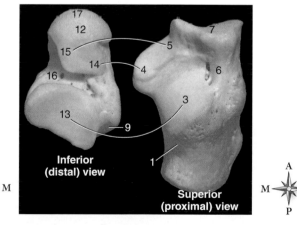

Figure 2-87 *Right ankle (subtalar joint; articular surfaces).*

1 Calcaneus
2 Fibular trochlea of calcaneus
3 Articular surface of talus for ankle joint
4 Medial tubercle of talus
5 Lateral tubercle of talus
6 Neck of talus
7 Head of talus
8 Navicular
9 Navicular tuberosity
10 Cuboid
11 Groove for fibularis longus
12 1st Cuneiform
13 2nd Cuneiform
14 3rd Cuneiform
15 Base of 1st metatarsal
16 Body (shaft) of 1st metatarsal
17 Head of 1st metatarsal
18 Tuberosity of base of 5th metatarsal
19 Base of 5th metatarsal
20 Body (shaft) of 5th metatarsal
21 Head of 5th metatarsal
22 Sesamoid bone of big toe
23 Proximal phalanx of big toe
24 Distal phalanx of big toe
25 Base of proximal phalanx of 2nd toe
26 Body (shaft) of proximal phalanx of 2nd toe
27 Head of proximal phalanx of 2nd toe
28 Middle phalanx of 3rd toe
29 Distal phalanx of 4th toe

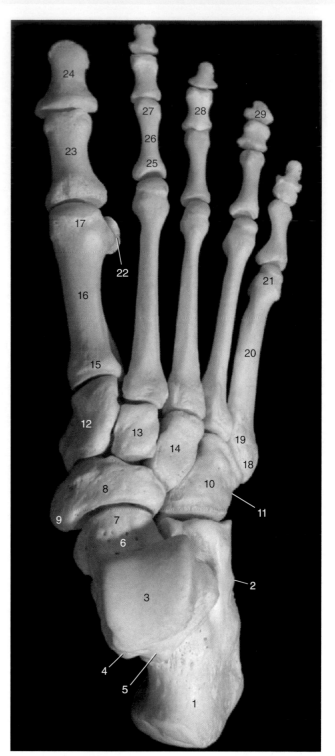

Figure 2-88 *Right foot (dorsal view).*

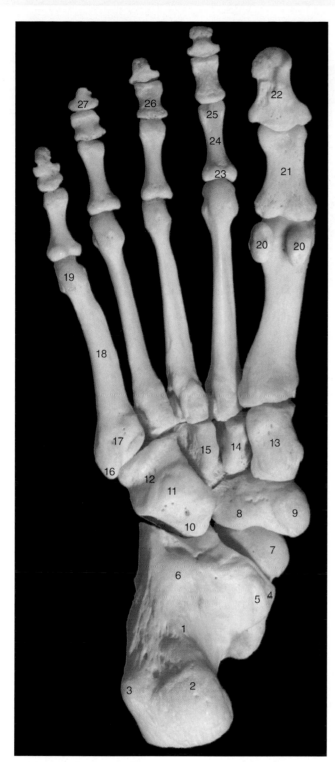

Figure 2-89 *Right foot (plantar view).*

1 Calcaneus
2 Medial process of calcaneal tuberosity
3 Lateral process of calcaneal tuberosity
4 Sustentaculum tali of calcaneus
5 Groove for distal tendon of flexor hallucis longus muscle
   (on sustentaculum tali)
6 Anterior tubercle of calcaneus
7 Head of talus
8 Navicular
9 Navicular tuberosity
10 Cuboid
11 Tuberosity of cuboid
12 Groove for distal tendon of fibularis longus muscle
13 1st Cuneiform
14 2nd Cuneiform
15 3rd Cuneiform
16 Tuberosity of base of 5th metatarsal
17 Base of 5th metatarsal
18 Body (shaft) of 5th metatarsal
19 Head of 5th metatarsal
20 Sesamoid bone of big toe
21 Proximal phalanx of big toe
22 Distal phalanx of big toe
23 Base of proximal phalanx of 2nd toe
24 Body (shaft) of proximal phalanx of 2nd toe
25 Head of proximal phalanx of 2nd toe
26 Middle phalanx of 3rd toe
27 Distal phalanx of 4th toe

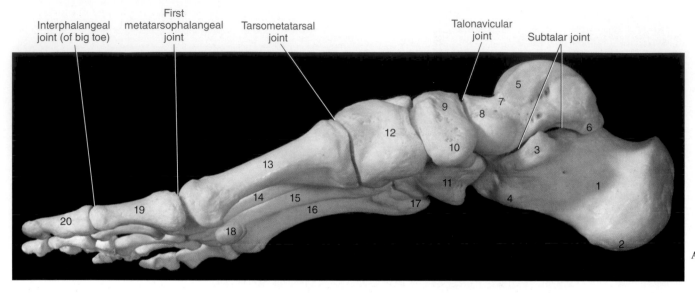

Figure 2-90 *Right foot (medial view).*

1 Calcaneus (medial surface)	7 Neck of talus	15 4th Metatarsal
2 Medial process of calcaneal tuberosity	8 Head of talus	16 5th Metatarsal
	9 Navicular	17 Tuberosity of base of 5th metatarsal
3 Sustentaculum tali of calcaneus	10 Navicular tuberosity	
4 Anterior tubercle of calcaneus	11 Cuboid	18 Sesamoid bone of big toe
5 Articular surface of talus for (medial malleolus of) ankle joint	12 1st Cuneiform	19 Proximal phalanx of big toe
	13 1st Metatarsal	20 Distal phalanx of big toe
6 Medial tubercle of talus	14 3rd Metatarsal	

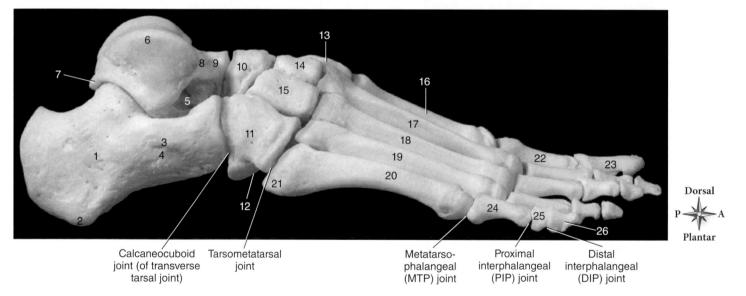

Figure 2-91 *Right foot (lateral view).*

1 Calcaneus (lateral surface)	7 Lateral tubercle of talus	17 2nd Metatarsal
2 Lateral process of calcaneal tuberosity	8 Neck of talus	18 3rd Metatarsal
	9 Head of talus	19 4th Metatarsal
3 Fibular trochlea	10 Navicular	20 5th Metatarsal
4 Groove for distal tendon of fibularis longus muscle	11 Cuboid	21 Tuberosity of base of 5th metatarsal
	12 Groove for distal tendon of fibularis longus muscle	22 Proximal phalanx of big toe
5 Tarsal sinus	13 1st Cuneiform	23 Distal phalanx of big toe
6 Articular surface of talus for (lateral malleolus of) ankle joint	14 2nd Cuneiform	24 Proximal phalanx of little toe
	15 3rd Cuneiform	25 Middle phalanx of little toe
	16 1st Metatarsal	26 Distal phalanx of little toe

Internal Anatomy

1 Left frontal sinus
2 Left ethmoidal air cells
3 Falx cerebri
4 Medial surface of right cerebral hemisphere
5 Anterior cerebral artery
6 Corpus callosum
7 Arachnoid granulations
8 Superior sagittal sinus
9 Tentorium cerebelli
10 Straight sinus
11 Cerebellum
12 Great cerebral vein
13 Midbrain
14 Pons
15 Fourth ventricle
16 Medulla oblongata
17 Margin of foramen magnum
18 Cerebellomedullary cistern (cisterna magna)
19 Posterior arch of atlas
20 Spinal cord
21 Intervertebral disk between axis and third cervical vertebra
22 Laryngopharynx
23 Inlet of larynx
24 Thyroid cartilage
25 Hyoid bone
26 Epiglottis
27 Vallecula
28 Oropharynx
29 Tongue
30 Mandible
31 Hard palate
32 Soft palate
33 Nasopharynx
34 Dens of axis
35 Anterior arch of atlas
36 Pharyngeal tonsil
37 Opening of auditory tube
38 Choana (posterior nasal aperture)
39 Nasal septum
40 Sphenoidal sinus
41 Pituitary gland
42 Optic chiasma

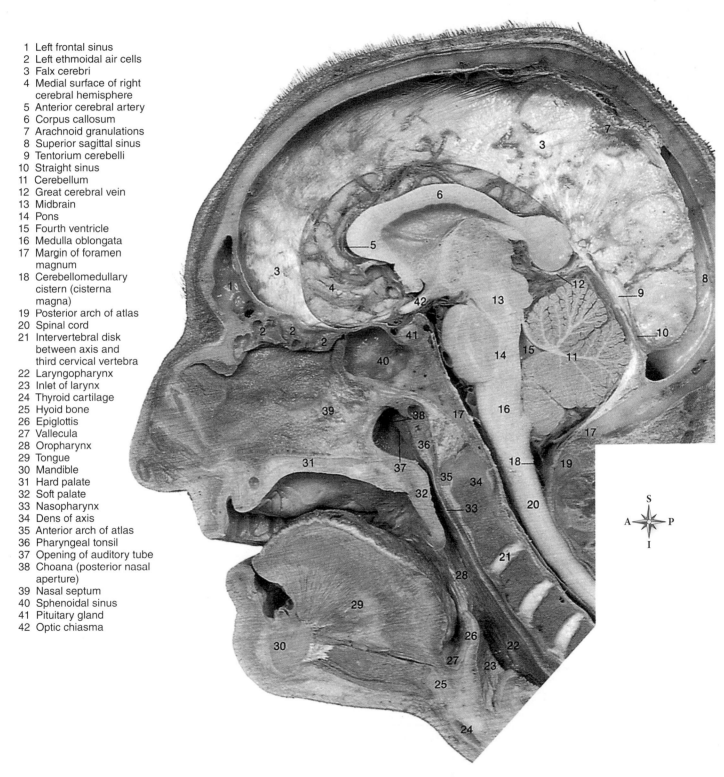

Figure 3-1 *Right half of the head, in sagittal section (lateral view).*

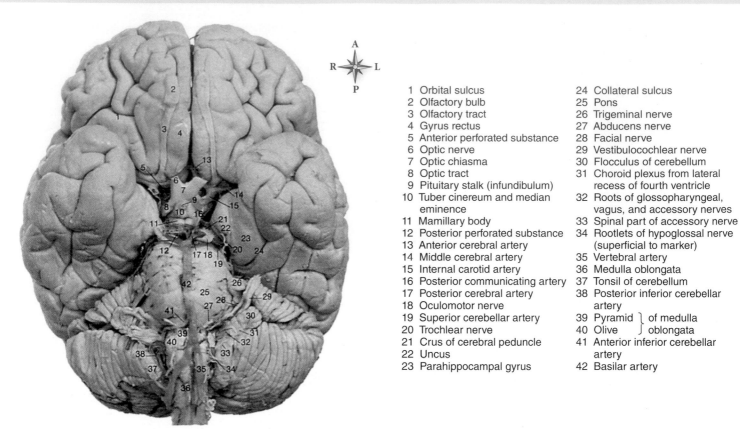

1 Orbital sulcus
2 Olfactory bulb
3 Olfactory tract
4 Gyrus rectus
5 Anterior perforated substance
6 Optic nerve
7 Optic chiasma
8 Optic tract
9 Pituitary stalk (infundibulum)
10 Tuber cinereum and median eminence
11 Mamillary body
12 Posterior perforated substance
13 Anterior cerebral artery
14 Middle cerebral artery
15 Internal carotid artery
16 Posterior communicating artery
17 Posterior cerebral artery
18 Oculomotor nerve
19 Superior cerebellar artery
20 Trochlear nerve
21 Crus of cerebral peduncle
22 Uncus
23 Parahippocampal gyrus

24 Collateral sulcus
25 Pons
26 Trigeminal nerve
27 Abducens nerve
28 Facial nerve
29 Vestibulocochlear nerve
30 Flocculus of cerebellum
31 Choroid plexus from lateral recess of fourth ventricle
32 Roots of glossopharyngeal, vagus, and accessory nerves
33 Spinal part of accessory nerve
34 Rootlets of hypoglossal nerve (superficial to marker)
35 Vertebral artery
36 Medulla oblongata
37 Tonsil of cerebellum
38 Posterior inferior cerebellar artery
39 Pyramid ⎫ of medulla
40 Olive ⎭ oblongata
41 Anterior inferior cerebellar artery
42 Basilar artery

Figure 3-2 *Brain (inferior view).*

1 Anterior cerebral artery
2 Rostrum ⎫
3 Genu ⎬ of corpus callosum
4 Body ⎭
5 Cingulate gyrus
6 Precentral gyrus
7 Central sulcus
8 Postcentral gyrus
9 Parietooccipital sulcus
10 Calcarine sulcus
11 Lingual gyrus
12 Cerebellum
13 Medulla oblongata
14 Median aperture of fourth ventricle
15 Fourth ventricle
16 Pons
17 Basilar artery
18 Tegmentum ⎫
19 Aqueduct ⎬ of midbrain
20 Inferior colliculus ⎪
21 Superior colliculus ⎭
22 Posterior commissure
23 Pineal body
24 Suprapineal recess
25 Great cerebral vein
26 Splenium of corpus callosum

27 Fornix
28 Cut edge of septum pellucidum
29 Body of lateral ventricle
30 Thalamus
31 Interthalamic connection
32 Hypothalamic sulcus
33 Hypothalamus
34 Posterior perforated substance
35 Mamillary body
36 Tuber cinereum and median eminence
37 Infundibular recess (base of pituitary stalk)
38 Optic chiasma
39 Supraoptic recess
40 Lamina terminalis
41 Anterior commissure
42 Anterior column of fornix
43 Interventricular foramen and choroid plexus

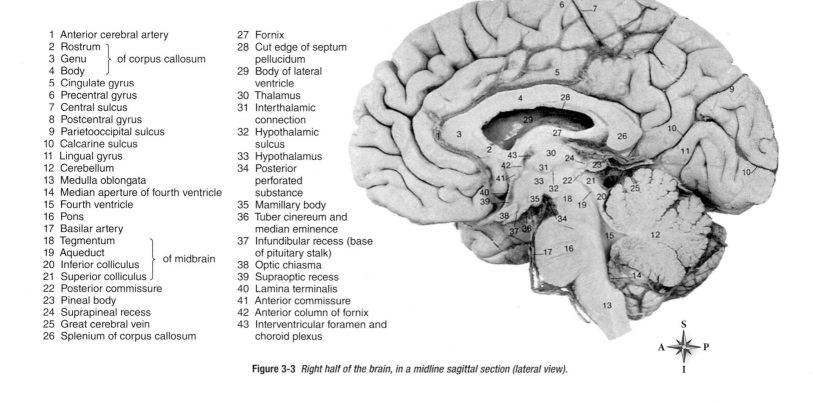

Figure 3-3 *Right half of the brain, in a midline sagittal section (lateral view).*

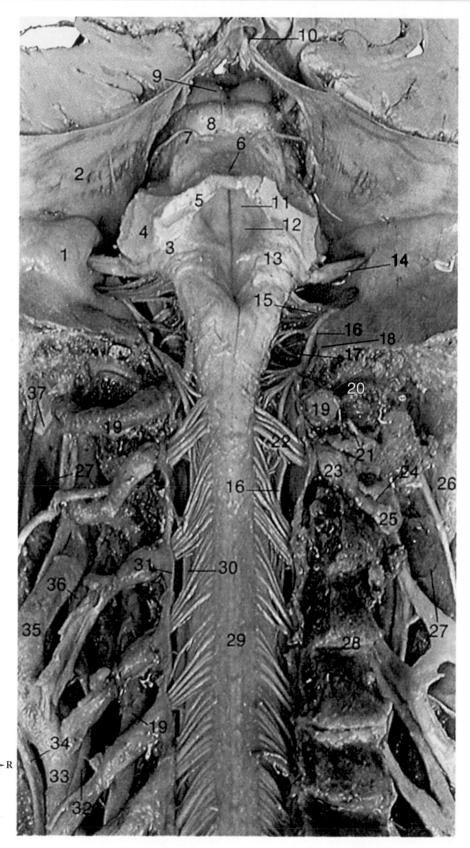

1 Petrous part of temporal bone
2 Tentorium cerebelli
3 Inferior ⎫
4 Middle ⎬ cerebellar
5 Superior ⎭ peduncle
6 Superior medullary velum
7 Trochlear nerve
8 Inferior ⎫
9 Superior ⎬ colliculus
10 Straight sinus
11 Medial eminence
12 Facial colliculus
13 Medullary striae
14 Facial and vestibulocochlear nerves and
   internal acoustic meatus
15 Glossopharyngeal, vagus, and
   accessory nerves and jugular foramen
16 Spinal part of accessory nerve
17 Rootlets of hypoglossal nerve and
   hypoglossal canal
18 Margin of foramen magnum
19 Vertebral artery
20 Lateral mass of atlas
21 Ventral ramus of first cervical nerve
22 Dorsal rootlets ⎫
23 Dorsal root ganglion ⎬ of second
24 Ventral ramus ⎬ cervical nerve
25 Dorsal ramus ⎭
26 Posterior belly of digastric muscle
27 Internal jugular vein
28 Zygapophyseal joint
29 Spinal cord
30 Denticulate ligament
31 Dura mater
32 Sympathetic trunk
33 Common carotid artery
34 Vagus nerve
35 Internal carotid artery
36 Superior cervical sympathetic ganglion
37 Hypoglossal nerve

Figure 3-4 *Brainstem and upper part of the spinal cord (posterior view).*

1 Conus medullaris of spinal cord
2 Cauda equina
3 Dura mater
4 Superior articular process of
  third lumbar vertebra
5 Filum terminale
6 Roots of fifth lumbar nerve
7 Fourth lumbar intervertebral disk
8 Pedicle of fifth lumbar vertebra
9 Dorsal root ganglion of fifth
  lumbar nerve
10 Fifth lumbar (lumbosacral) vertebra
11 Dural sheath of first sacral nerve
  roots
12 Lateral part of sacrum
13 Second sacral vertebra

Figure 3-5 *Vertebral column, lumbar and sacral regions (posterior view).*

1 Lateral cord
2 Posterior cord
3 Medial cord
4 Pectoralis minor and lateral pectoral nerve
5 Musculocutaneous nerve
6 Axillary nerve
7 Lateral root of median nerve
8 Radial nerve
9 Medial root of median nerve
10 Upper subscapular nerves
11 Thoracodorsal nerve
12 Lower subscapular nerve
13 Medial cutaneous nerve of arm
14 Ulnar nerve
15 Medial cutaneous nerve of forearm
16 Intercostobrachial nerve
17 Subscapularis
18 Teres major
19 Latissimus dorsi
20 Long head of triceps
21 Lateral head of triceps
22 Medial head of triceps
23 Radial nerve branches to triceps
24 Median nerve
25 Coracobrachialis
26 Biceps
27 Deltoid

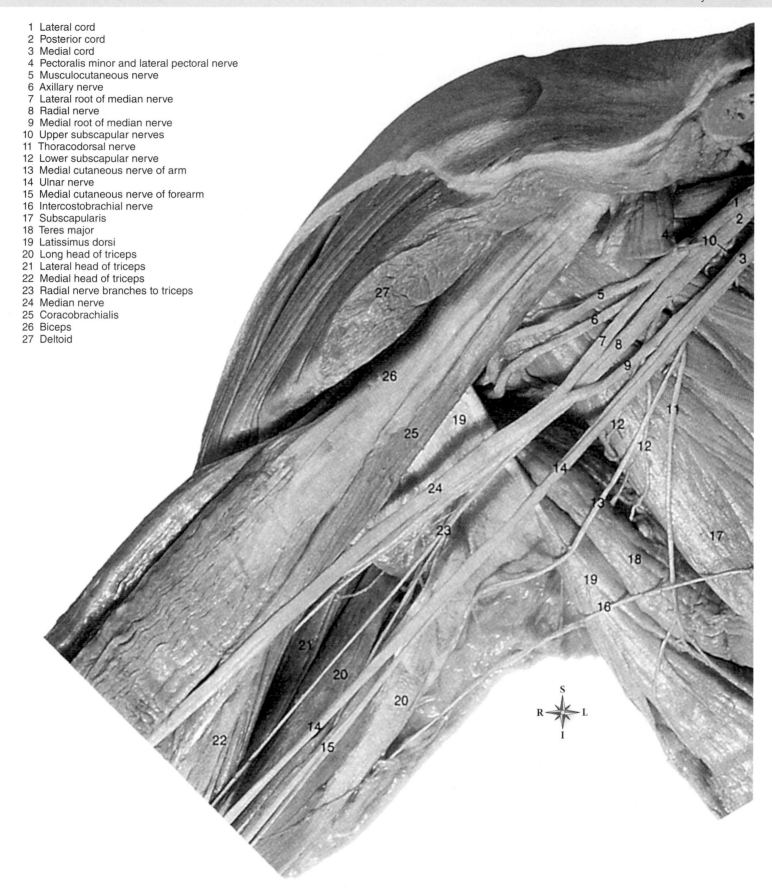

Figure 3-6 *Right brachial plexus and branches (anterior view).*

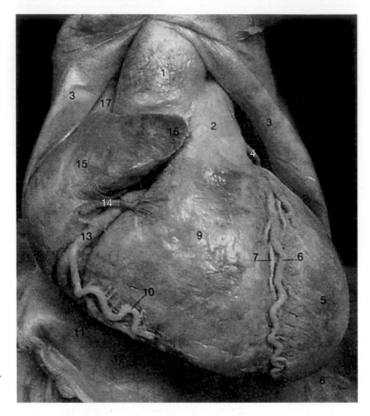

1 Ascending aorta
2 Pulmonary trunk
3 Serous pericardium overlying fibrous pericardium (turned laterally)
4 Auricle of left atrium
5 Left ventricle
6 Anterior interventricular branch of left coronary artery
7 Great cardiac vein
8 Diaphragm
9 Right ventricle
10 Marginal branch of right coronary artery
11 Small cardiac vein
12 Pericardium fused with tendon of diaphragm
13 Right coronary artery
14 Anterior cardiac vein
15 Right atrium
16 Auricle of right atrium
17 Superior vena cava

Figure 3-7 *Heart and pericardium (anterior view).*

1 Arch of cricoid cartilage
2 Isthmus   ⎱ of thyroid
3 Lateral lobe ⎰ gland
4 Trachea
5 Inferior thyroid veins
6 Left common carotid artery
7 Left vagus nerve
8 Internal jugular vein
9 Subclavian vein
10 Thoracic duct
11 Internal thoracic vein
12 Internal thoracic artery
13 Phrenic nerve
14 Parietal pleura (cut edge) over lung
15 Left brachiocephalic vein
16 A thymic artery
17 Thymic veins
18 Thymus
19 Superior vena cava
20 Right brachiocephalic vein
21 First rib
22 Brachiocephalic trunk
23 Right common carotid artery
24 Right subclavian artery
25 Right recurrent laryngeal nerve
26 Right vagus nerve
27 Unusual cervical tributary of 20
28 Vertebral vein
29 Thyrocervical trunk
30 Suprascapular artery
31 Scalenus anterior
32 Upper trunk of brachial plexus
33 Superficial cervical artery
34 Ascending cervical artery
35 Inferior thyroid artery
36 Sympathetic trunk

Figure 3-8 *Thoracic inlet and mediastinum (anterior view).*

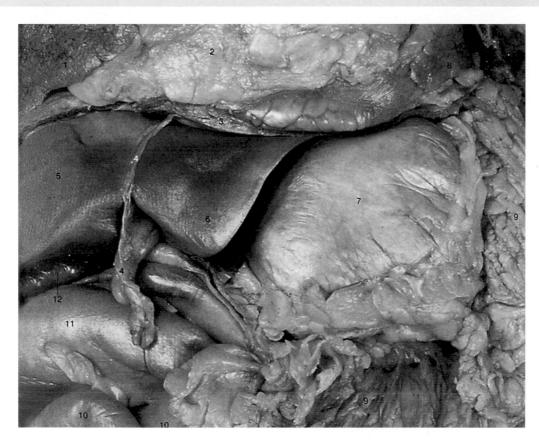

1 Inferior lobe of right lung
2 Pericardial fat
3 Diaphragm
4 Falciform ligament
5 Right lobe of liver
6 Left lobe of liver
7 Stomach
8 Inferior lobe of left lung
9 Greater omentum
10 Small intestine
11 Transverse colon
12 Gallbladder

Figure 3-9 *Upper abdominal viscera (anterior view).*

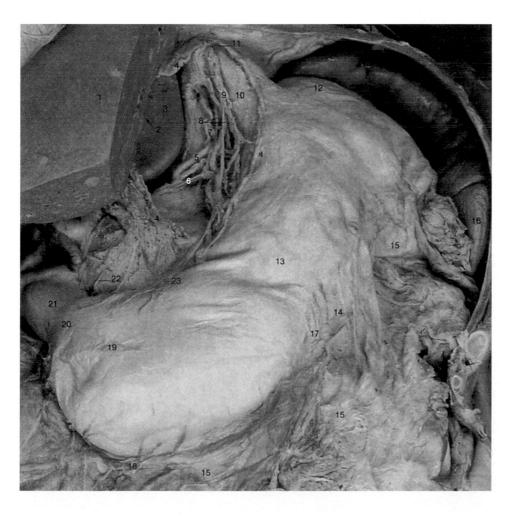

1 Right lobe of liver
2 Fissure of ligamentum venosum
3 Caudate lobe of liver
4 Lesser omentum (cut edge)
5 Left gastric artery
6 Left gastric vein
7 Posterior vagal trunk
8 Esophageal branches of left gastric vessels
9 Anterior vagal trunk
10 Esophagus
11 Esophageal opening in diaphragm
12 Fundus ⎤
13 Body ⎬ of stomach
14 Greater curvature ⎦
15 Greater omentum
16 Lower end of spleen
17 Branches of left gastroepiploic vessels
18 Right gastroepiploic vessels and branches
19 Pyloric part of stomach
20 Pylorus
21 Superior (first) part of duodenum
22 Right gastric artery
23 Lesser curvature

Figure 3-10 *Stomach, with vessels and vagus nerves (anterior view).*

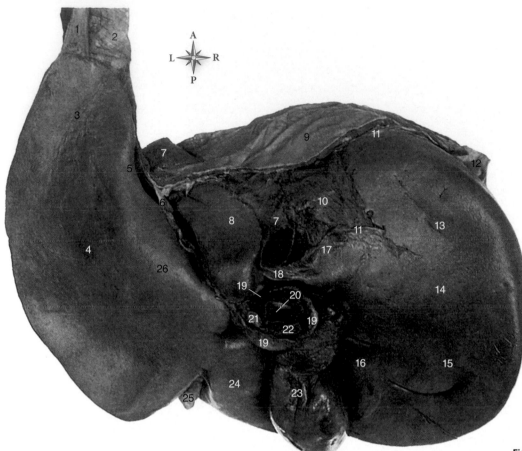

1 Left triangular ligament
2 Diaphragm
3 Left lobe
4 Gastric impression
5 Esophageal groove
6 Lesser omentum in fissure for
   ligamentum venosum
7 Inferior vena cava
8 Caudate lobe
9 Diaphragm on part of bare area
10 Bare area
11 Inferior layer of coronary ligament
12 Right triangular ligament
13 Renal impression
14 Right lobe
15 Colic impression
16 Duodenal impression
17 Suprarenal impression
18 Caudate process
19 Right free margin of lesser
   omentum in porta hepatis
20 Portal vein
21 Hepatic artery
22 Common hepatic duct
23 Gallbladder
24 Quadrate lobe
25 Ligamentum teres and falciform
   ligament in fissure for legamentum
   teres
26 Omental tuberosity

Figure 3-11 *Liver (from above and behind).*

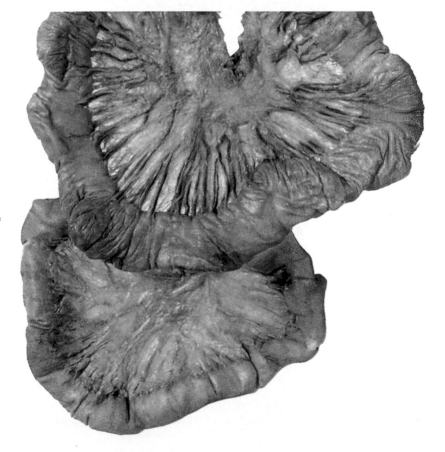

Figure 3-12 *Small intestine.* Coil of
typical jejunum, coil of typical ileum.

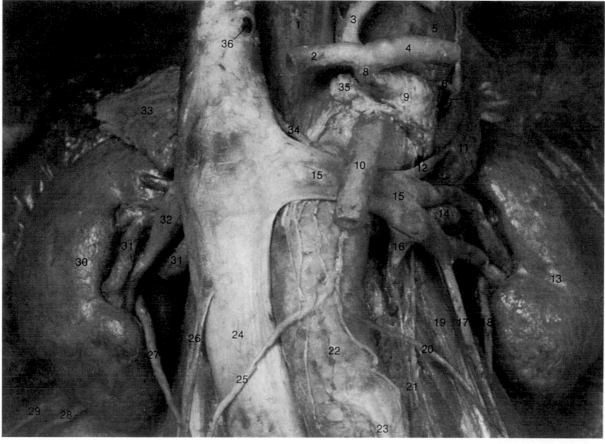

1 Right crus of diaphragm
2 Common hepatic artery
3 Left gastric artery
4 Splenic artery
5 Left crus of diaphragm
6 Left inferior phrenic artery
7 Left inferior phrenic vein
8 Celiac trunk
9 Left celiac ganglion
10 Superior mesenteric artery
11 Left suprarenal gland
12 Left suprarenal vein
13 Left kidney
14 Left renal artery
15 Left renal vein
16 Lumbar tributary of renal vein
17 Left gonadal vein
18 Left ureter

19 Left psoas major
20 Left gonadal artery
21 Left sympathetic trunk
22 Abdominal aorta and aortic plexus
23 Inferior mesenteric artery
24 Inferior vena cava
25 Right gonadal artery
26 Right gonadal vein
27 Right ureter
28 Right ilioinguinal nerve
29 Right iliohypogastric nerve
30 Right kidney
31 Right renal artery
32 Right renal vein
33 Right suprarenal gland
34 Right inferior phrenic artery
35 Right celiac ganglion
36 A hepatic vein

Figure 3-13 *Kidneys and suprarenal glands (anterior view).*

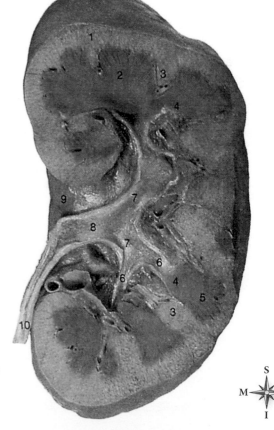

1 Cortex
2 Medulla
3 Renal column
4 Renal papilla
5 Medullary pyramid

6 Minor calyx
7 Major calyx
8 Renal pelvis
9 Hilum
10 Ureter

Figure 3-14 *Kidney.* Internal structure in frontal section.

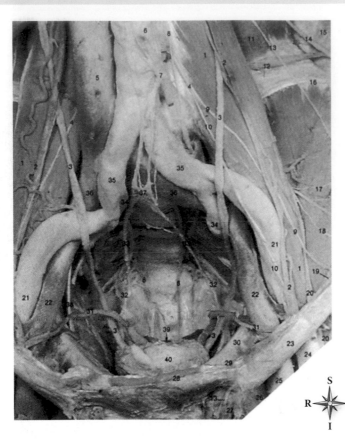

1 Psoas major
2 Testicular vessels
3 Ureter
4 Genitofemoral nerve
5 Inferior vena cava
6 Aorta and aortic plexus
7 Inferior mesenteric artery and plexus
8 Sympathetic trunk and ganglia
9 Femoral ⎫ branch of
10 Genital ⎭ genitofemoral nerve
11 Quadratus lumborum
12 Fourth lumbar artery
13 Ilioinguinal nerve
14 Iliohypogastric nerve
15 Lumbar part of thoracolumbar fascia
16 Iliolumbar ligament
17 Iliacus and branches from femoral nerve and iliolumbar artery
18 Lateral femoral cutaneous nerve arising from femoral nerve
19 Deep circumflex iliac artery

20 Femoral nerve
21 External iliac artery
22 External iliac vein
23 Inguinal ligament
24 Femoral artery
25 Femoral vein
26 Position of femoral canal
27 Spermatic cord
28 Rectus abdominis
29 Lacunar ligament
30 Pectineal ligament
31 Ductus deferens
32 Inferior hypogastric (pelvic) plexus and pelvic splanchnic nerves
33 Hypogastric nerve
34 Internal iliac artery
35 Common iliac artery
36 Common iliac vein
37 Superior hypogastric plexus
38 Obturator nerve and vessels
39 Rectum (cut edge)
40 Bladder

Figure 3-15 *Posterior abdominal and pelvic walls (anterior view).*

1 Rectus abdominis
2 Extraperitoneal fat
3 Sigmoid colon
4 Promontory of sacrum
5 Rectum
6 Coccyx
7 Anococcygeal body
8 External anal sphincter
9 Anal canal with anal columns of mucous membrane
10 Perineal body
11 Ductus deferens
12 Epididymis
13 Testis
14 Spongy part of urethra and corpus spongiosum
15 Corpus cavernosum
16 Bulbospongiosus
17 Perineal membrane
18 Sphincter urethrae
19 Membranous part of urethra
20 Pubic symphysis
21 Prostate
22 Prostatic part of urethra
23 Seminal colliculus
24 Bristle in ejaculatory duct
25 Internal urethral orifice
26 Bladder
27 Bristle passing up into right ureteral orifice
28 Rectovesical pouch
29 Puborectalis fibers of levator ani

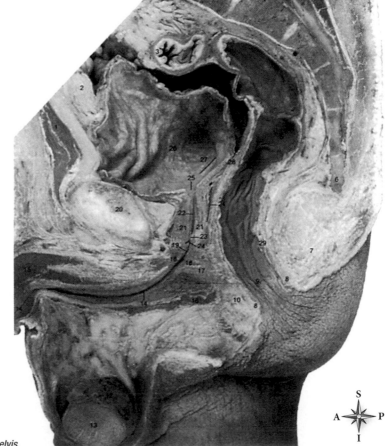

Figure 3-16 *Right half of a midline sagittal section of the male pelvis.*

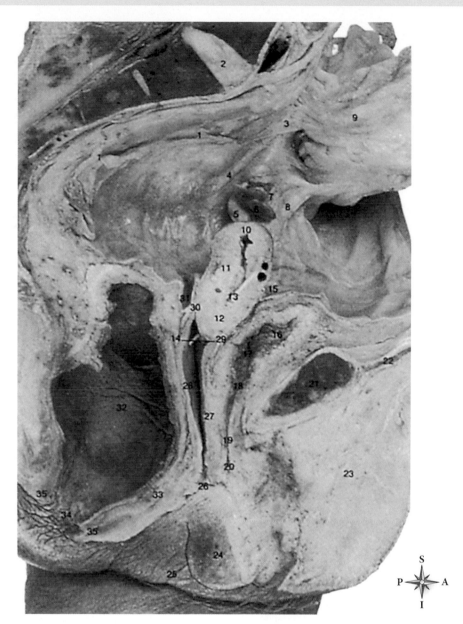

1 Line of attachment of right limb of sigmoid mesocolon	17 Marker in left ureteral orifice
2 Fifth lumbar intervertebral disk	18 Internal urethral orifice
3 Apex of sigmoid mesocolon	19 Urethra
4 Ureter underlying peritoneum	20 External urethral orifice
5 Ovary	21 Pubic symphysis
6 Uterine tube	22 Rectus abdominis (turned forward)
7 Suspensory ligament of ovary containing ovarian vessels	23 Fat of mons pubis
8 Left limb of sigmoid mesocolon overlying external iliac vessels	24 Labium minus
9 Sigmoid colon (reflected to left and upward)	25 Labium majus

1 Line of attachment of right limb of sigmoid mesocolon
2 Fifth lumbar intervertebral disk
3 Apex of sigmoid mesocolon
4 Ureter underlying peritoneum
5 Ovary
6 Uterine tube
7 Suspensory ligament of ovary containing ovarian vessels
8 Left limb of sigmoid mesocolon overlying external iliac vessels
9 Sigmoid colon (reflected to left and upward)
10 Fundus ⎤
11 Body ⎬ of uterus
12 Cervix ⎦
13 Marker in internal os
14 Marker in external os
15 Vesicouterine pouch
16 Bladder

17 Marker in left ureteral orifice
18 Internal urethral orifice
19 Urethra
20 External urethral orifice
21 Pubic symphysis
22 Rectus abdominis (turned forward)
23 Fat of mons pubis
24 Labium minus
25 Labium majus
26 Vestibule ⎤
27 Anterior wall ⎥
28 Posterior wall ⎬ of vagina
29 Anterior fornix ⎥
30 Posterior fornix ⎦
31 Recto-uterine pouch
32 Rectum
33 Perineal body
34 Anal canal
35 External anal sphincter

Figure 3-17 *Left half of a midline sagittal section of the female pelvis.*

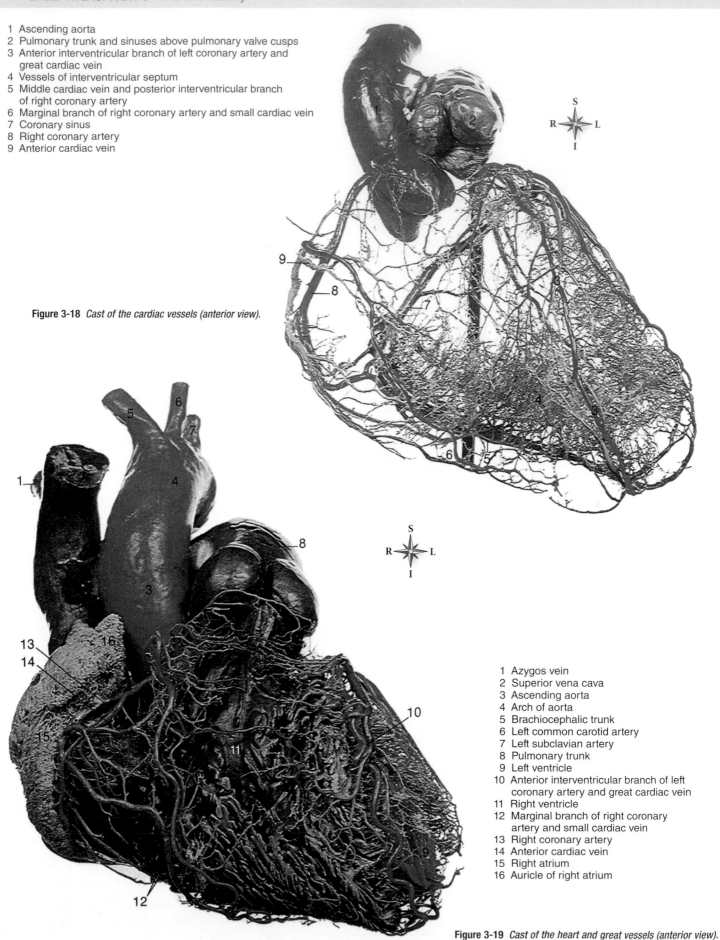

1 Ascending aorta
2 Pulmonary trunk and sinuses above pulmonary valve cusps
3 Anterior interventricular branch of left coronary artery and great cardiac vein
4 Vessels of interventricular septum
5 Middle cardiac vein and posterior interventricular branch of right coronary artery
6 Marginal branch of right coronary artery and small cardiac vein
7 Coronary sinus
8 Right coronary artery
9 Anterior cardiac vein

Figure 3-18 *Cast of the cardiac vessels (anterior view).*

1 Azygos vein
2 Superior vena cava
3 Ascending aorta
4 Arch of aorta
5 Brachiocephalic trunk
6 Left common carotid artery
7 Left subclavian artery
8 Pulmonary trunk
9 Left ventricle
10 Anterior interventricular branch of left coronary artery and great cardiac vein
11 Right ventricle
12 Marginal branch of right coronary artery and small cardiac vein
13 Right coronary artery
14 Anterior cardiac vein
15 Right atrium
16 Auricle of right atrium

Figure 3-19 *Cast of the heart and great vessels (anterior view).*

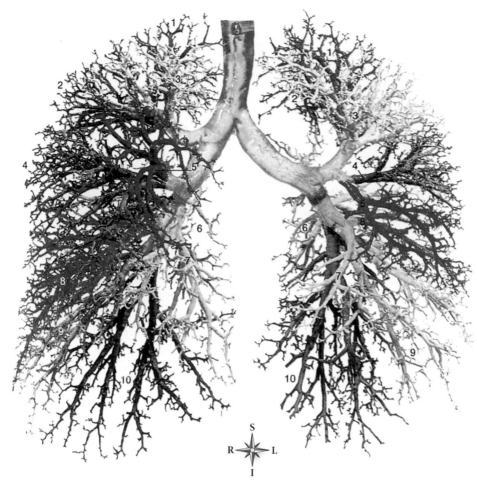

RIGHT LUNG
Superior lobe
 1 Apical
 2 Posterior
 3 Anterior

Middle lobe
 4 Lateral
 5 Medial

Inferior lobe
 6 Apical (superior)
 7 Medial basal
 8 Anterior basal
 9 Lateral basal
10 Posterior basal

LEFT LUNG
Superior lobe
 1 Apical
 2 Posterior
 3 Anterior
 4 Superior lingular
 5 Inferior lingular

Inferior lobe
 6 Apical (superior)
 7 Medial basal (cardiac)
 8 Anterior basal
 9 Lateral basal
10 Posterior basal

Figure 3-20 *Cast of the bronchial tree (anterior view).*

 1 Right branch of portal vein and hepatic
   artery and right hepatic duct
 2 Gallbladder
 3 Bile duct
 4 Hepatic artery
 5 Portal vein
 6 Left branch of portal vein and hepatic
   artery and left hepatic duct
 7 Left gastric artery
 8 Left gastric vein
 9 Splenic artery
10 Splenic vein
11 Short gastric vessels
12 Left gastroepiploic vessels
13 Vessels of left kidney
14 Pancreatic duct
15 Duodenojejunal flexure
16 Superior mesenteric artery
17 Superior mesenteric vein
18 Horizontal (third) part of duodenum
19 Right gastroepiploic vessels
20 Pyloric canal
21 Pylorus
22 Superior (first) part of duodenum
23 Right gastric vessels
24 Branches of superior and inferior
   pancreaticoduodenal vessels
25 Descending (second) part of duodenum
26 Vessels of right kidney

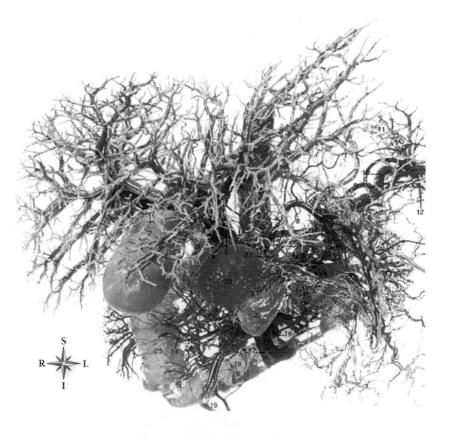

Figure 3-21 *Cast of the duodenum, liver, biliary tract,
and associated vessels (anterior view).*

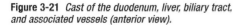

1 Right renal vein
2 Right suprarenal vein
3 Inferior vena cava
4 Aorta
5 Celiac trunk
6 Superior mesenteric
   artery

7 Left renal vein
8 Left suprarenal veins
9 Left renal artery
10 Accessory renal
   arteries
11 Right renal artery

Figure 3-22 *Cast of the kidneys and great vessels (anterior view).*

Cross-Sectional Anatomy

1 Arytenoid cartilage
2 Claustrum
3 Common carotid artery
4 Ethmoidal air cells
5 Head of caudate nucleus
6 Internal capsule of cerebrum
7 Internal jugular vein
8 Lamina of vertebra
9 Lateral rectus muscle
10 Lens
11 Lentiform nucleus
12 Levator scapulae muscle
13 Ligamentum nuchae
14 Longus colli muscle
15 Medial rectus muscle
16 Nasal cavity
17 Optic canal
18 Optic chiasma
19 Optic nerve
20 Optic radiation
21 Orbital fat
22 Piriform fossa, pharynx
23 Platysma muscle
24 Scalenus anterior muscle
25 Scalenus medius and scalenus posterior muscles
26 Semispinalis capitis muscle
27 Spinal cord
28 Spinalis muscle
29 Splenius capitis muscle
30 Sternocleidomastoid muscle
31 Superior sagittal sinus
32 Temporal lobe, cerebrum
33 Temporalis muscle
34 Thalamus
35 Thyroid cartilage
36 Thyroid gland, lateral lobe
37 Trapezius muscle
38 Vertebral artery in transverse foramen
39 Vertebral body
40 Vertebral canal
41 Vocal cord
42 Zygomatic bone

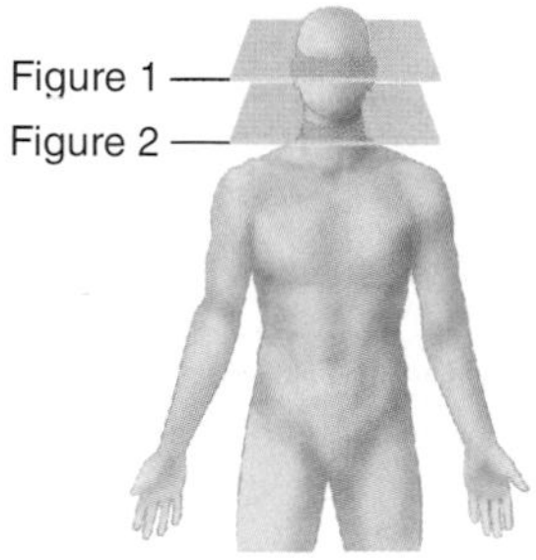

Figure 1
Figure 2

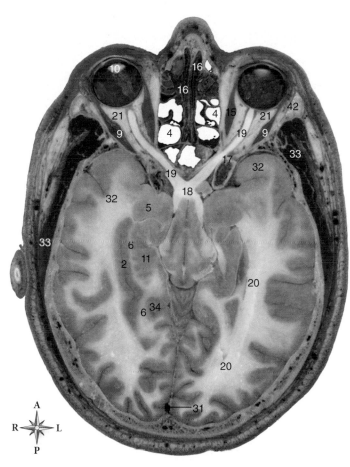

Figure 4-1 *Head and neck (inferior view).* Cross section at level of optic chiasma.

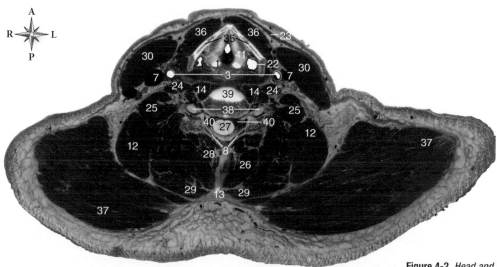

Figure 4-2 *Head and neck (inferior view).* Cross section at level of vocal cords.

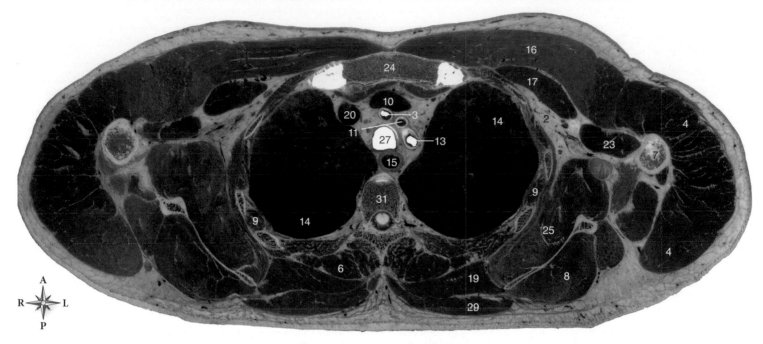

Figure 4-3 *Thorax (inferior view).* Cross section at T2 vertebral level.

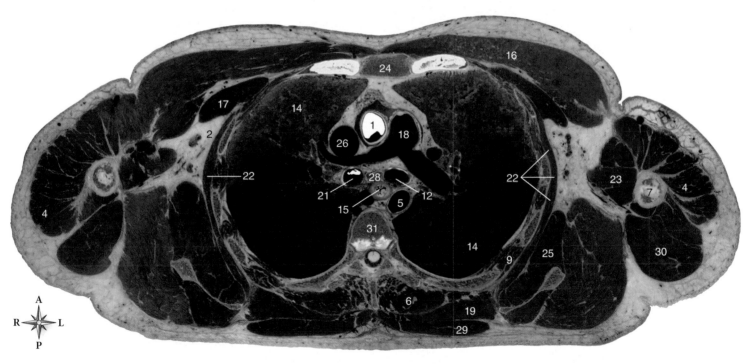

1 Ascending aorta
2 Axillary fat with brachial plexus
3 Brachiocephalic artery
4 Deltoid muscle
5 Descending aorta
6 Erector spinae muscle
7 Humerus
8 Infraspinatus muscle
9 Intercostal muscles
10 Left brachiocephalic vein
11 Left common carotid artery

12 Left main bronchus
13 Left subclavian artery
14 Lung
15 Esophagus
16 Pectoralis major muscle
17 Pectoralis minor muscle
18 Pulmonary trunk
19 Rhomboid major muscle
20 Right brachiocephalic vein
21 Right main bronchus
22 Serratus anterior muscle

23 Short head of biceps brachii
24 Sternal marrow
25 Subscapularis muscle
26 Superior vena cava
27 Trachea
28 Tracheobronchial lymph
29 Trapezius muscle
30 Triceps muscle
31 Vertebral body

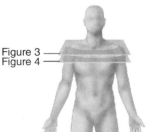

Figure 3
Figure 4

Figure 4-4 *Thorax (inferior view).* Cross section at T4 vertebral level.

1 Aorta
2 Body of pancreas
3 Descending colon
4 Diaphragm
5 Duodenum
6 Erector spinae muscle
7 External oblique muscle
8 Gallbladder
9 Head of pancreas
10 Inferior vena cava
11 Intercostal muscle
12 Intervertebral disk
13 Kidney
14 Latissimus dorsi muscle
15 Left crus of diaphragm
16 Left renal vein
17 Linea alba
18 Linea semilunaris
19 Liver
20 Perirenal fat
21 Portal vein
22 Psoas muscle
23 Quadratus abdominis muscle
24 Rectus abdominis muscle
25 Rib
26 Right crus of diaphragm
27 Small intestine
28 Spinal cord
29 Spleen
30 Splenic artery and vein
31 Stomach
32 Superior mesenteric vessels
33 Tail of pancreas
34 Transverse colon
35 Vertebral body

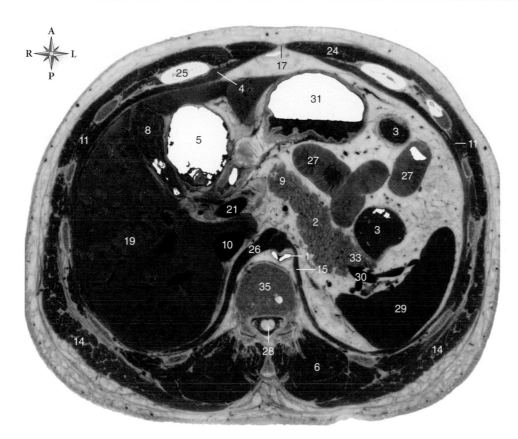

Figure 4-5 *Abdomen (inferior view).* Cross section at L1 vertebral level.

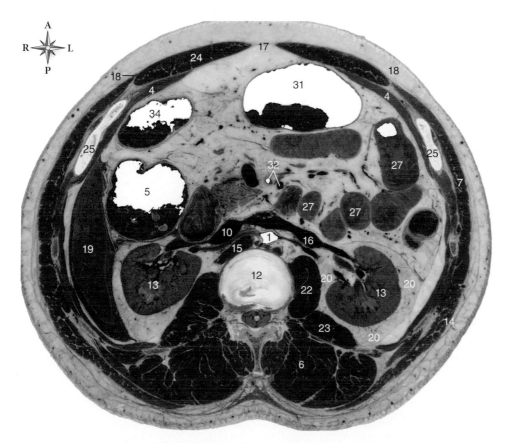

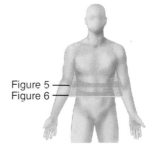

Figure 5
Figure 6

Figure 4-6 *Abdomen (inferior view).* Cross section at L2 vertebral level.

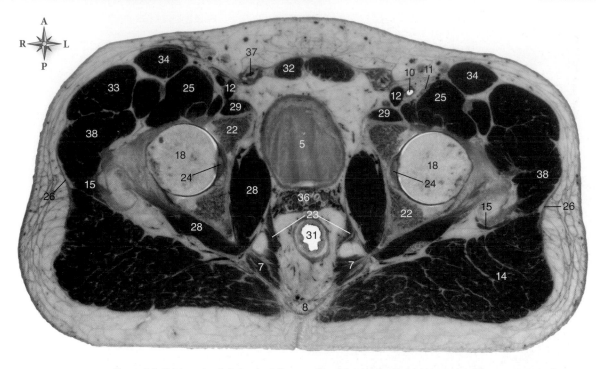

1 Adductor brevis muscle
2 Adductor longus muscle
3 Adductor magnus muscle
4 Anal canal
5 Bladder
6 Bulb of penis
7 Coccygeus part of levator ani muscle
8 Coccyx
9 Crus of penis
10 Femoral artery
11 Femoral nerve
12 Femoral vein
13 Femur
14 Gluteus maximus muscle
15 Gluteus minimus muscle
16 Great saphenous vein
17 Hamstring origin
18 Head of femur
19 Ischiocavernosus muscle
20 Ischial tuberosity
21 Ischioanal fossa
22 Ischium
23 Levator ani muscle
24 Ligament of head of femur
25 Iliopsoas muscle
26 Iliotibial tract
27 Obturator externus muscle
28 Obturator internus muscle
29 Pectineus muscle
30 Quadratus femoris muscle
31 Rectum
32 Rectus abdominis muscle
33 Rectus femoris muscle
34 Sartorius muscle
35 Sciatic nerve
36 Seminal vesicles
37 Spermatic cord
38 Tensor fasciae latae muscle
39 Vastus intermedius muscle
40 Vastus lateralis muscle

Figure 4-7 *Pelvic region (inferior view).* Cross section at level of the hip joint in a male pelvis.

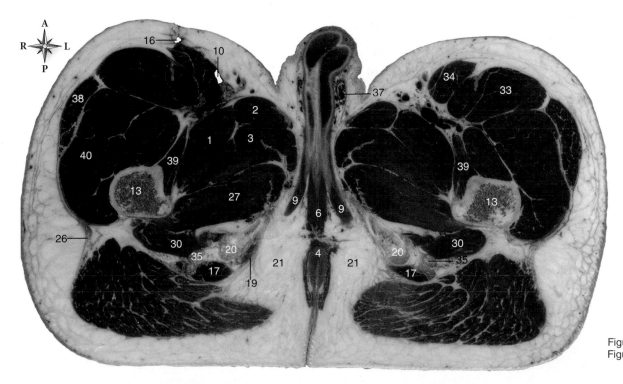

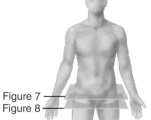

Figure 7
Figure 8

Figure 4-8 *Pelvic region (inferior view).* Cross section at level of the upper thigh in a male pelvis.

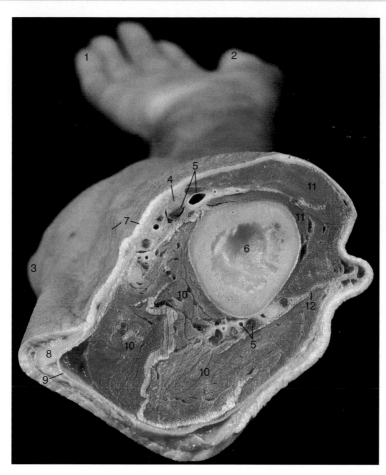

1 Digit 5 (little finger)
2 Digit 1 (thumb)
3 Medial epicondyle (surface bump)
4 Nerve
5 Blood vessels
6 Humerus
7 Skin
8 Superficial fascia
9 Deep fascia
10 Muscles of the posterior compartment
11 Muscles of the anterior compartment
12 Lateral intermuscular septum

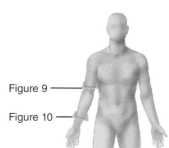

Figure 9
Figure 10

Figure 4-9 *Upper arm.* Cross section proximal to the elbow.

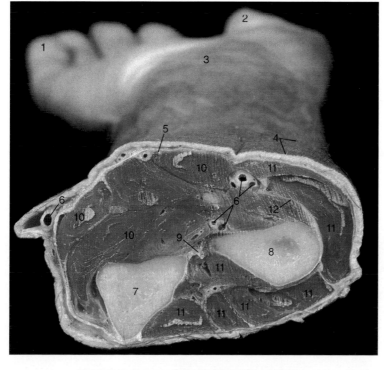

1 Digit 5 (little finger)
2 Digit 1 (thumb)
3 Carpus (wrist)
4 Skin
5 Deep fascia
6 Blood vessels
7 Ulna
8 Radius
9 Interosseous ligament
10 Muscles of the anterior compartment
11 Muscles of the posterior compartment
12 Intermuscular septum

Figure 4-10 *Lower arm.* Cross section distal to the elbow.

1 Epidermis
2 Dermis
3 Superficial fascia
4 Digital artery
5 Tendon (flexor digitorum superficialis)
6 Tendon (flexor digitorum profundis)
7 Tendon sheath
8 Proximal phalanx
9 Tendon (extensor expansion)
10 Digital nerve
11 Digital vein

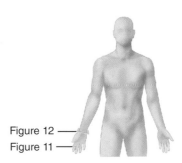

Figure 12
Figure 11

Figure 4-11 *Second digit (index finger).* Cross section at the proximal phalanx.

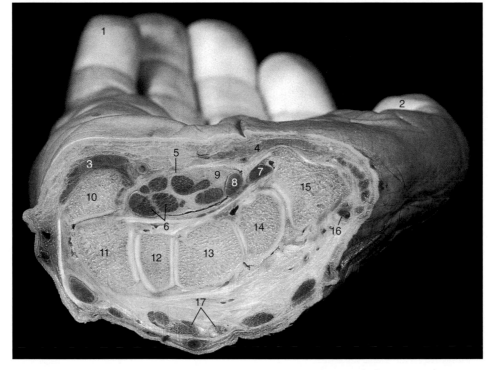

1 Digit 5 (little finger)
2 Digit 1 (thumb)
3 Hypothenar muscles
4 Thenar muscles
5 Tendon sheath (of carpal tunnel)
6 Tendons (digital flexor)
7 Flexor carpi radialis
8 Flexor pollicis longus
9 Median nerve
10 Pisiform
11 Triquetral
12 Hamate
13 Capitate
14 Trapezoid
15 Trapezium
16 Radial artery
17 Tendons (digital extensor)

Figure 4-12 *Carpus (wrist).* Cross section showing the carpal tunnel.

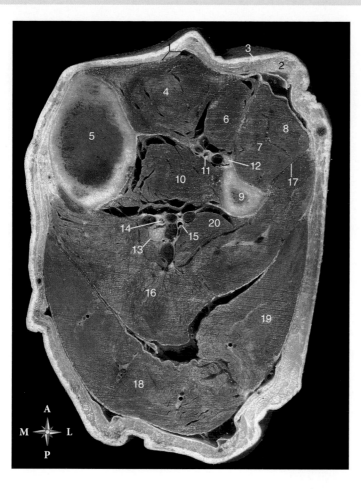

1 Deep fascia
2 Superficial fascia
3 Skin
4 Tibialis anterior
5 Tibia
6 Extensor digitorum longus
7 Peroneus brevis
8 Peroneus longus
9 Fibula
10 Tibialis posterior
11 Anterior tibial artery
12 Deep peroneal nerve
13 Tibial nerve
14 Posterior tibial artery
15 Peroneal artery
16 Soleus
17 Posterior peroneal intermuscular septum
18 Gastrocnemius (medial head)
19 Gastrocnemius (lateral head)
20 Flexor hallucis longus

Figure 4-13 *Leg.* Cross section showing bones and muscles below the knee.

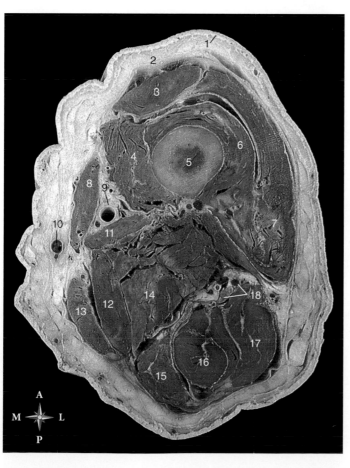

1 Skin
2 Superficial fascia
3 Rectus femoris
4 Vastus medialis
5 Femur
6 Vastus intermedius
7 Vastus lateralis
8 Sartorius
9 Femoral artery
10 Great saphenous vein
11 Adductor longus
12 Adductor brevis
13 Gracilis
14 Adductor magnus
15 Semimembranosus
16 Semitendinosus
17 Biceps femoris
18 Branches of sciatic nerve

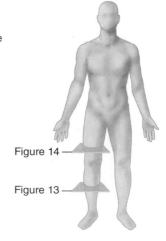

Figure 14 —
Figure 13 —

Figure 4-14 *Thigh.* Cross section showing major muscles above the knee.

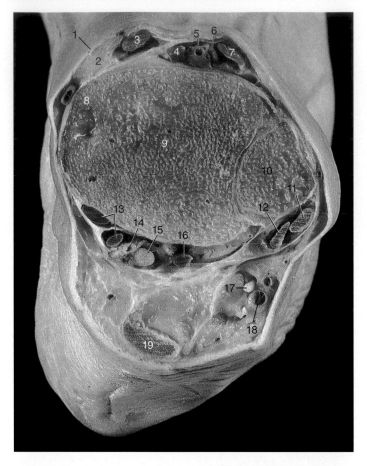

1 Skin
2 Superficial fascia
3 Tendon (tibialis anterior)
4 Tendon (extensor hallucis longus)
5 Anterior tibial artery
6 Deep peroneal nerve
7 Tendon (extensor digitorum longus)
8 Great saphenous vein
9 Tibia
10 Lateral malleolus of fibula
11 Tendon (peroneus longus)
12 Tendon (peroneus brevis)
13 Tendons (tibialis posterior, flexor digitorum longus)
14 Posterior tibial artery
15 Posterior tibial nerve
16 Tendon (flexor hallucis longus)
17 Sural nerve
18 Short saphenous vein
19 Tendon (calcaneal or Achilles)

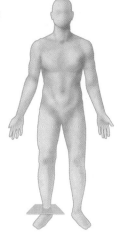

Figure 4-15 *Lower leg.* Cross section just above the ankle joint cavity, showing relations of structures.

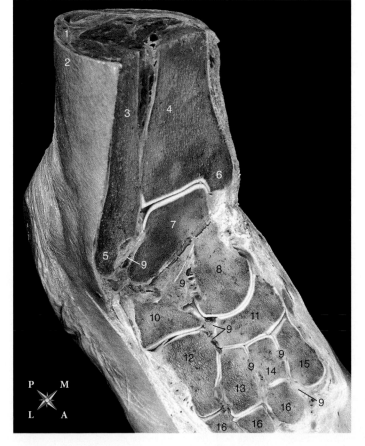

1 Superficial fascia
2 Skin
3 Fibula
4 Tibia
5 Lateral malleolus of fibula
6 Medial malleolus of tibia
7 Trochlea of talus
8 Head of talus
9 Interosseous ligaments
10 Calcaneus
11 Navicular
12 Cuboid
13 Third cuneiform
14 Second cuneiform
15 First cuneiform
16 Base of metatarsal

Figure 4-16 *Ankle and foot.* Combined cross section through anterior part of tarsal region and frontal (coronal) section through the lower leg and tarsal region.

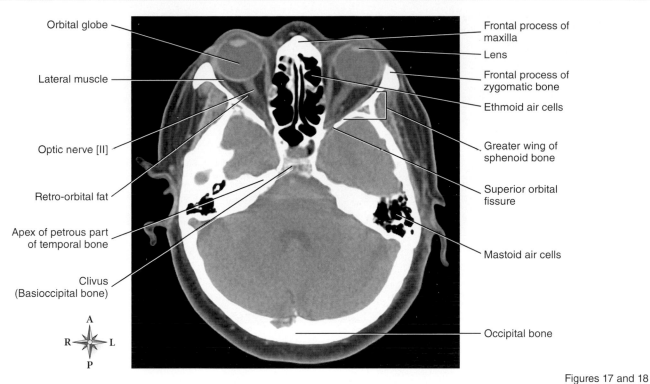

Orbital globe

Lateral muscle

Optic nerve [II]

Retro-orbital fat

Apex of petrous part
of temporal bone

Clivus
(Basioccipital bone)

Frontal process of
maxilla

Lens

Frontal process of
zygomatic bone

Ethmoid air cells

Greater wing of
sphenoid bone

Superior orbital
fissure

Mastoid air cells

Occipital bone

A
R — L
P

Figure 4-17 *Head.* Computed tomography (CT) scan showing a cross section of
the head at the level of the eye orbit.

Figures 17 and 18

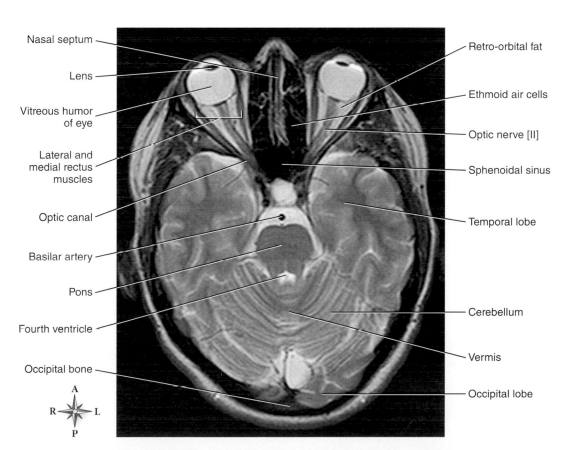

Nasal septum

Lens

Vitreous humor
of eye

Lateral and
medial rectus
muscles

Optic canal

Basilar artery

Pons

Fourth ventricle

Occipital bone

Retro-orbital fat

Ethmoid air cells

Optic nerve [II]

Sphenoidal sinus

Temporal lobe

Cerebellum

Vermis

Occipital lobe

A
R — L
P

Figure 4-18 *Head.* Magnetic resonance imaging (MRI) scan showing a cross section of the head at the level of the
eye orbit. Compare the similar cross section above (Figure 4-17) and note that visualization of individual structures
varies depending on the imaging technology used to produce the image. (See *Medical Imaging of the Body* online at
A&P Connect.)

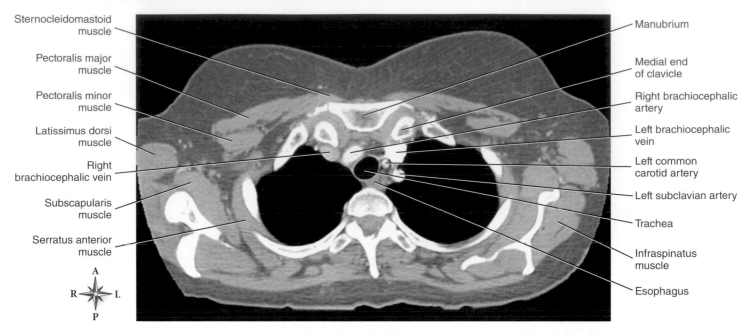

Sternocleidomastoid muscle

Pectoralis major muscle

Pectoralis minor muscle

Latissimus dorsi muscle

Right brachiocephalic vein

Subscapularis muscle

Serratus anterior muscle

Manubrium

Medial end of clavicle

Right brachiocephalic artery

Left brachiocephalic vein

Left common carotid artery

Left subclavian artery

Trachea

Infraspinatus muscle

Esophagus

Figure 4-19 *Thorax.* CT scan showing a cross section of the chest wall and structures of the mediastinum.

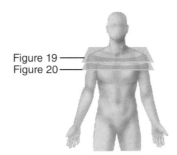

Figure 19
Figure 20

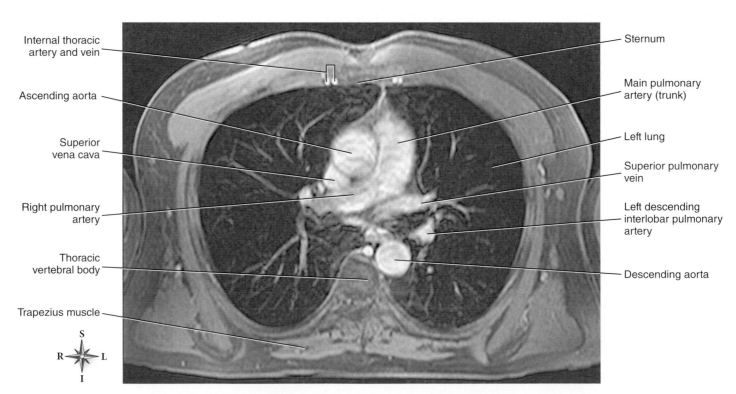

Internal thoracic artery and vein

Ascending aorta

Superior vena cava

Right pulmonary artery

Thoracic vertebral body

Trapezius muscle

Sternum

Main pulmonary artery (trunk)

Left lung

Superior pulmonary vein

Left descending interlobar pulmonary artery

Descending aorta

Figure 4-20 *Thorax.* MRI scan showing a cross section of the chest.

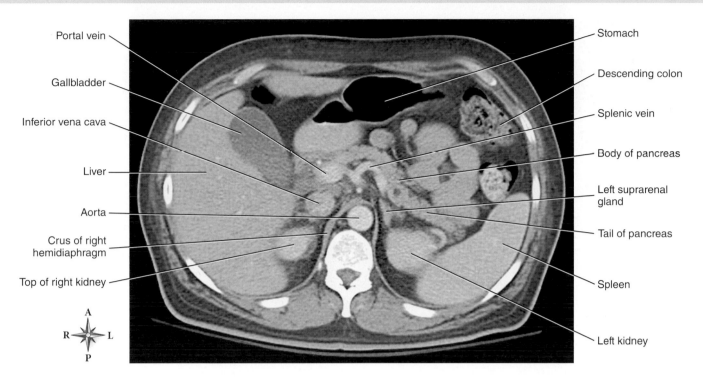

Portal vein

Gallbladder

Inferior vena cava

Liver

Aorta

Crus of right hemidiaphragm

Top of right kidney

Stomach

Descending colon

Splenic vein

Body of pancreas

Left suprarenal gland

Tail of pancreas

Spleen

Left kidney

Figure 4-21 *Abdomen.* CT scan showing a cross section of the upper abdominal wall and internal abdominal organs.

Figure 21
Figure 22

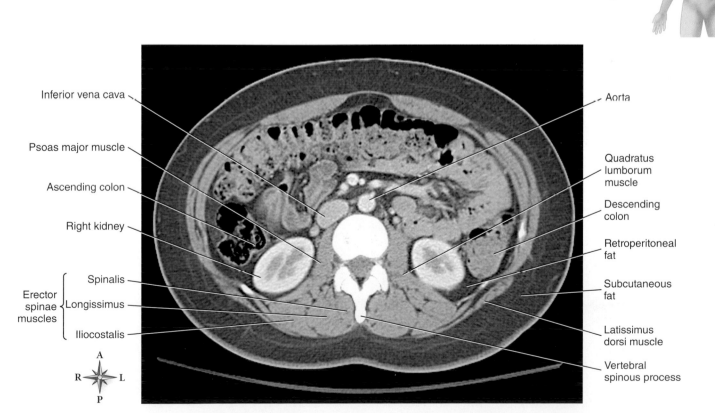

Inferior vena cava

Psoas major muscle

Ascending colon

Right kidney

Erector spinae muscles {
Spinalis
Longissimus
Iliocostalis
}

Aorta

Quadratus lumborum muscle

Descending colon

Retroperitoneal fat

Subcutaneous fat

Latissimus dorsi muscle

Vertebral spinous process

Figure 4-22 *Abdomen.* CT scan showing a cross section of the lower abdominal contents and wall. Note the clear definition of the muscles of the back.

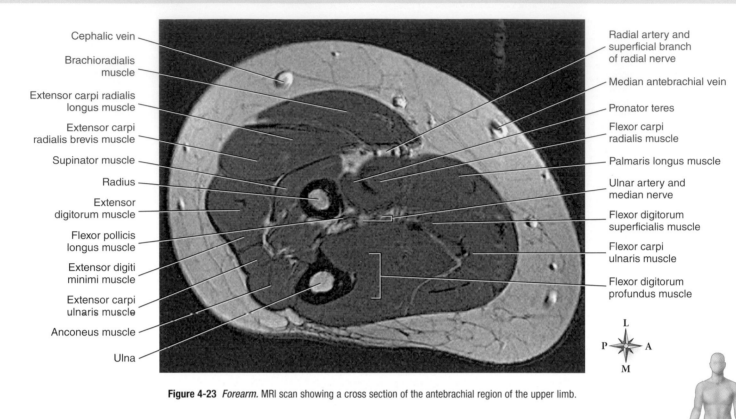

Cephalic vein

Brachioradialis muscle

Extensor carpi radialis longus muscle

Extensor carpi radialis brevis muscle

Supinator muscle

Radius

Extensor digitorum muscle

Flexor pollicis longus muscle

Extensor digiti minimi muscle

Extensor carpi ulnaris muscle

Anconeus muscle

Ulna

Radial artery and superficial branch of radial nerve

Median antebrachial vein

Pronator teres

Flexor carpi radialis muscle

Palmaris longus muscle

Ulnar artery and median nerve

Flexor digitorum superficialis muscle

Flexor carpi ulnaris muscle

Flexor digitorum profundus muscle

Figure 4-23 *Forearm.* MRI scan showing a cross section of the antebrachial region of the upper limb.

Figure 23 —

Figure 24 —

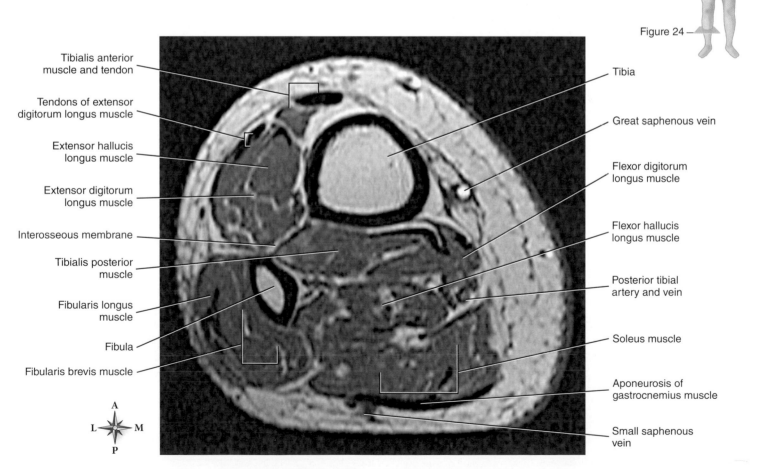

Tibialis anterior muscle and tendon

Tendons of extensor digitorum longus muscle

Extensor hallucis longus muscle

Extensor digitorum longus muscle

Interosseous membrane

Tibialis posterior muscle

Fibularis longus muscle

Fibula

Fibularis brevis muscle

Tibia

Great saphenous vein

Flexor digitorum longus muscle

Flexor hallucis longus muscle

Posterior tibial artery and vein

Soleus muscle

Aponeurosis of gastrocnemius muscle

Small saphenous vein

Figure 4-24 *Leg.* MRI scan showing a cross section near the calf of the leg.

Histology

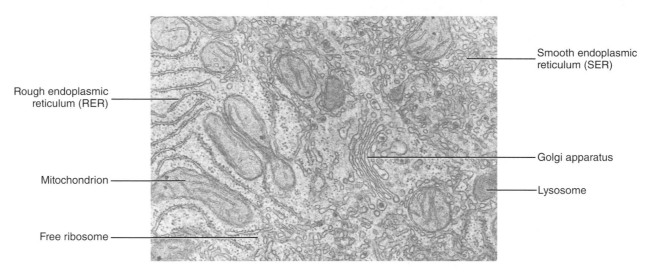

Figure 5-1 *Electron micrograph of a thin section of a liver cell showing organelles.*

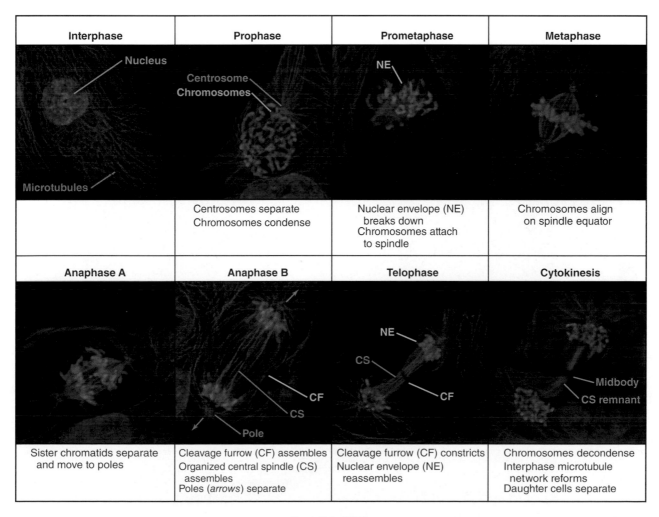

Interphase	Prophase	Prometaphase	Metaphase
	Centrosomes separate Chromosomes condense	Nuclear envelope (NE) breaks down Chromosomes attach to spindle	Chromosomes align on spindle equator

Anaphase A	Anaphase B	Telophase	Cytokinesis

Sister chromatids separate
and move to poles
	Cleavage furrow (CF) assembles
	Organized central spindle (CS)
	  assembles
	Poles (*arrows*) separate
		Cleavage furrow (CF) constricts
		Nuclear envelope (NE)
		  reassembles
			Chromosomes decondense
			Interphase microtubule
			  network reforms
			Daughter cells separate

Figure 5-2 *Mitosis.*

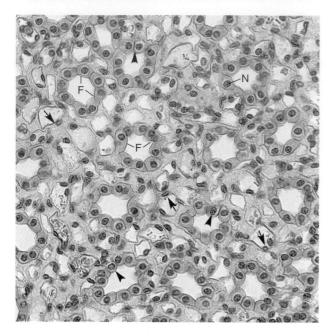

Figure 5-3 *Simple squamous epithelium* (arrows) *and simple cuboidal epithelium* (arrowheads) (×270). *F,* Free edge; *N,* nucleus.

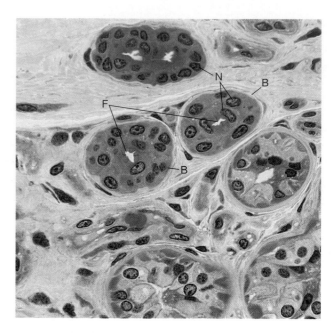

Figure 5-4 *Stratified cuboidal epithelium* (×509). *B,* Basement membrane; *F,* free edge; *N,* nucleus.

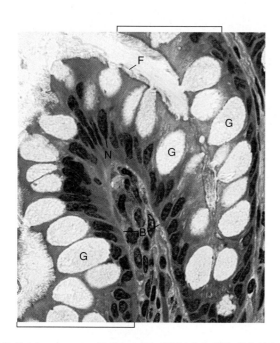

Figure 5-5 *Simple columnar epithelium with goblet cells* (×540). *G,* Goblet cell; *N,* nucleus; *B,* basement membrane; *F,* free edge.

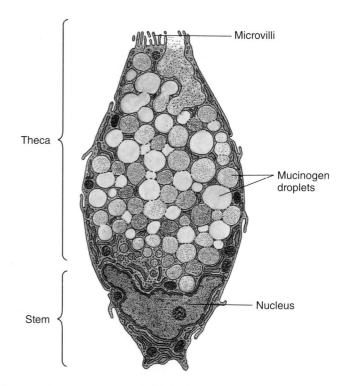

Figure 5-6 *Drawing of a goblet cell.* Schematic diagram of the ultrastructure of a goblet cell illustrating the tightly packed secretory granules of the theca.

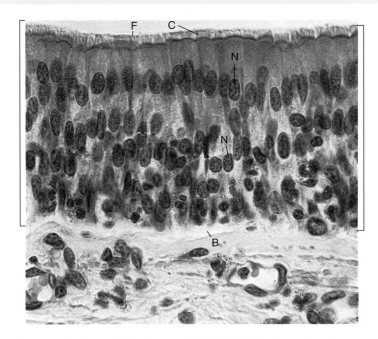

Figure 5-7 *Pseudostratified columnar epithelium* (×540). *B,* Basement membrane; *C,* cilia; *F,* free edge; *N,* nucleus.

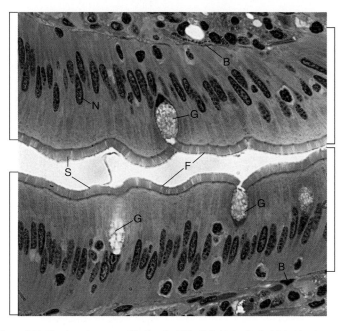

Figure 5-8 *Simple columnar epithelium* (×540). *G,* Goblet cell; *S,* striated border; *N,* nucleus; *F,* free edge; *B,* basement membrane.

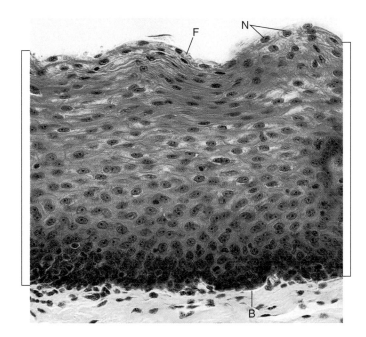

Figure 5-9 *Stratified squamous epithelium (nonkeratinized)* (×509). *B,* Basement membrane; *F,* free edge; *N,* nucleus.

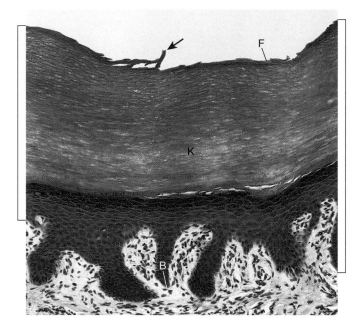

Figure 5-10 *Stratified squamous epithelium (keratinized)* (×125). *Arrow,* Flaking off of dead cells; *K,* keratinized layer; *F,* free edge; *B,* basement membrane.

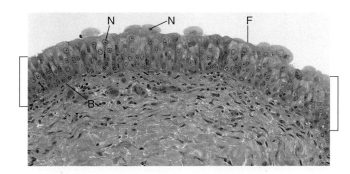

Figure 5-11 *Transitional epithelium* (×125). *B,* Basement membrane; *F,* free edge; *N,* nucleus.

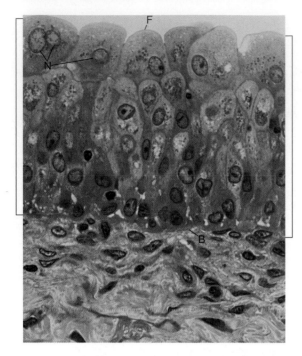

Figure 5-12 *Transitional epithelium* (×540). *B,* Basement membrane; *F,* free edge; *N,* nucleus.

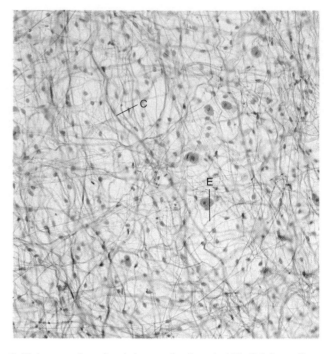

Figure 5-13 *Loose, ordinary (areolar) connective tissue* (×132). *C,* Collagen fiber; *E,* elastin fiber.

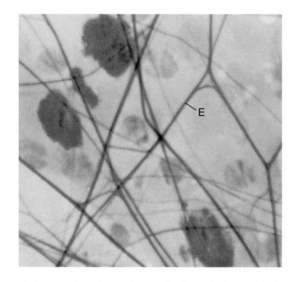

Figure 5-14 *Loose, ordinary (areolar) connective tissue* (high power). *E,* Elastin fiber.

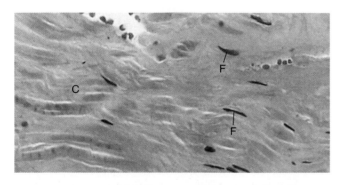

Figure 5-16 *Dense fibrous connective tissue. C,* Bundles of collagen fibers; *F,* fibroblasts.

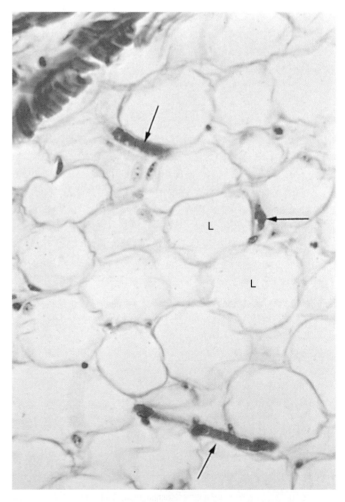

Figure 5-15 *Adipose tissue. L,* Lipid-storing vesicles; *arrows,* blood capillaries.

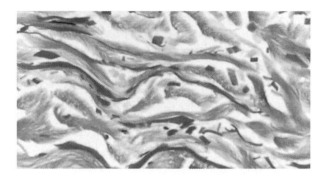

Figure 5-17 *Elastic fibrous connective tissue.*

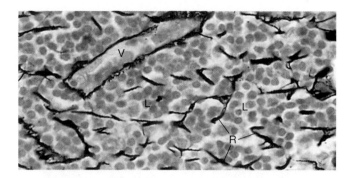

Figure 5-18 *Reticular connective tissue.* *L,* Lymphoid cells; *R,* reticular fibers; *V,* blood vessel.

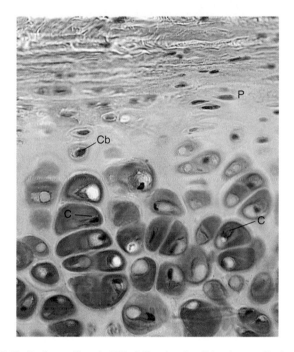

Figure 5-19 *Hyaline cartilage* (×270). *C,* Chondrocyte within a lacuna; *Cb,* chondroblast; *P,* perichondrium.

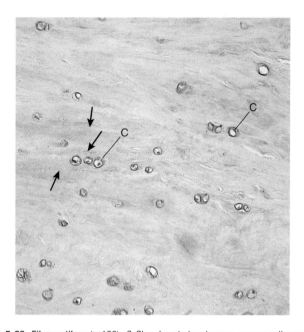

Figure 5-20 *Fibrocartilage* (×132). *C,* Chondrocyte in a lacuna; *arrows,* collagen fiber bundles.

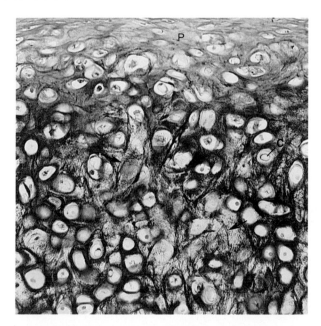

Figure 5-21 *Elastic cartilage* (×132). *P,* Perichondrium; *C,* chondrocyte in a lacuna; *arrows,* elastic fiber.

Figure 5-22 *Compact bone* (×270). *C,* Central (haversian) canal; *arrows,* canaliculi.

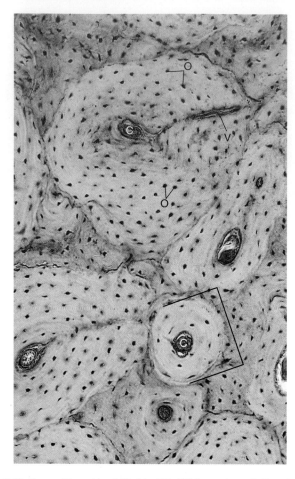

Figure 5-23 *Compact bone (decalcified)* (×162). *V,* Volkmann's canal; *C,* central canal; *O,* osteocyte; *bracket,* osteon (haversian system).

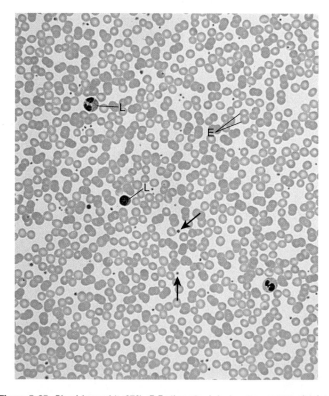

Figure 5-25 *Blood (smear)* (×270). *E,* Erythrocyte; *L,* leukocytes; *arrows,* platelets.

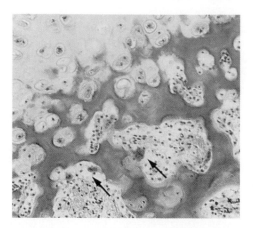

Figure 5-24 *Spongy (cancellous) bone tissue* (high power). *Arrows,* Osteoclasts.

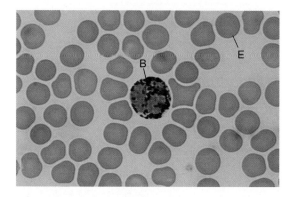

Figure 5-26 *Blood (smear showing basophil)* (×1325). *E,* Erythrocyte; *B,* basophil.

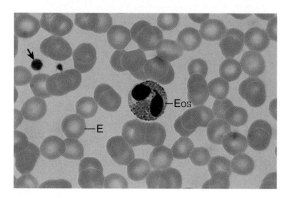

Figure 5-27 *Blood (smear showing eosinophil)* (×1325). *E,* Erythrocyte; *Eos,* eosino-phil; *arrow,* platelet.

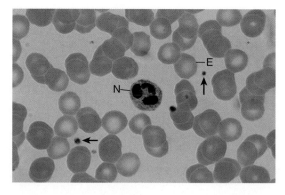

Figure 5-28 *Blood (smear showing neutrophil)* (×1325). *E,* Erythrocyte; *arrows,* platelets; *N,* neutrophil.

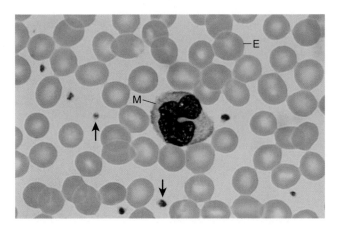

Figure 5-29 *Blood (smear showing monocyte)* (×1325). *E,* Erythrocyte; *M,* monocyte; *arrows,* platelets.

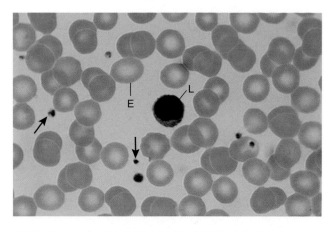

Figure 5-30 *Blood (smear showing lymphocyte)* (×1325). *E,* Erythrocyte; *arrows,* platelets; *L,* lymphocyte.

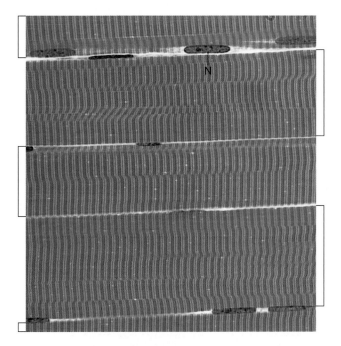

Figure 5-31 *Skeletal muscle (longitudinal section)* (×540). *N,* Nucleus.

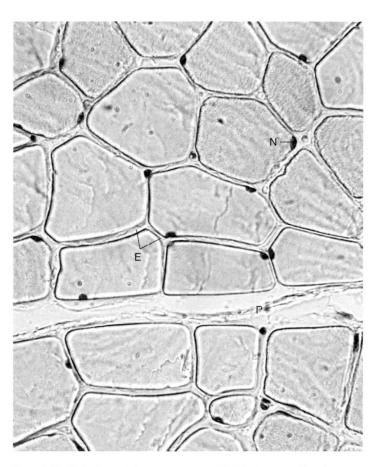

Figure 5-32 *Skeletal muscle (cross section)* (×540). *E,* Endomysium; *N,* nucleus; *P,* perimysium.

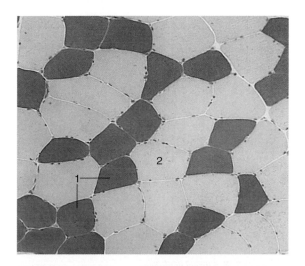

Figure 5-33 *Skeletal muscle (cross section).* *1,* Type 1 fibers; *2,* type 2 fibers.

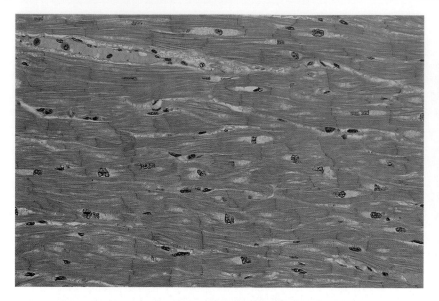

Figure 5-34 *Cardiac muscle showing branching (longitudinal section)* (×270)

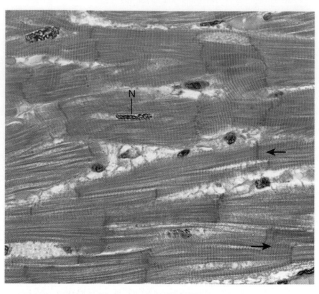

Figure 5-35 *Cardiac muscle (longitudinal section)* (×540). *N,* nucleus; *arrows,* intercalated disks.

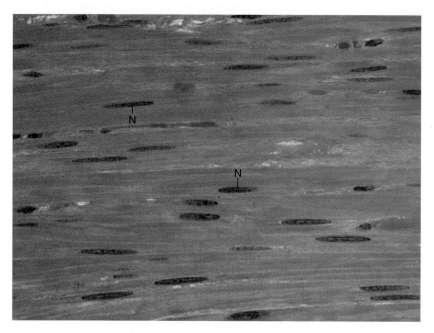

Figure 5-36 *Smooth muscle (longitudinal section)* (×540). *N,* Nuclei.

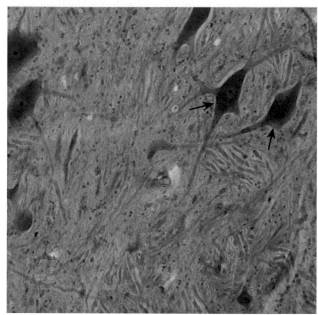

Figure 5-37 *Nerve tissue (spinal cord gray matter)* (×270). *Arrows,* Neurons.

Figure 5-38 *Nerve tissue (spinal cord smear). M,* Multipolar neuron; *N,* nuclei of glial cells.

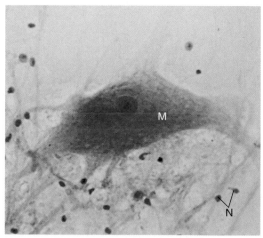

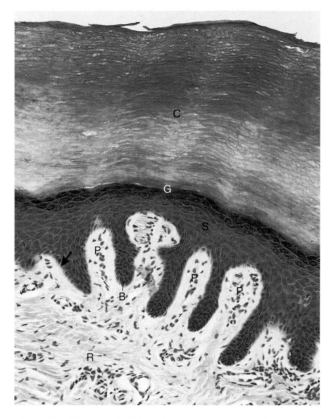

Figure 5-39 *Skin (thick)* (×132). Epidermis: *C,* Stratum corneum; *G,* stratum granulosum; *S,* stratum spinosum; *B,* stratum basale; *arrow,* dermal-epidermal junction. Dermis: *P,* Papilla (papillary region); *R,* reticular region.

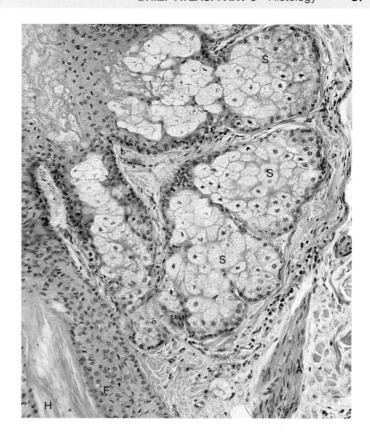

Figure 5-40 *Sebaceous gland* (×132). *S,* Sebaceous glands; *A,* arrector pili muscle; *H,* hair; *F,* hair follicle.

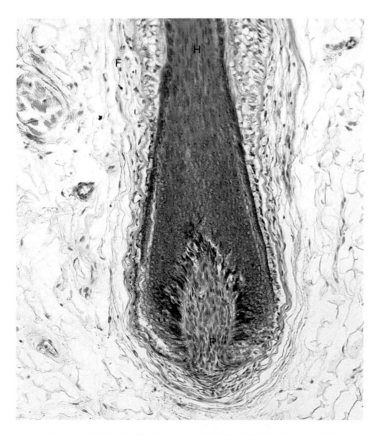

Figure 5-41 *Hair follicle (longitudinal section)* (×132). *F,* Follicle; *H,* hair; *P,* hair papilla.

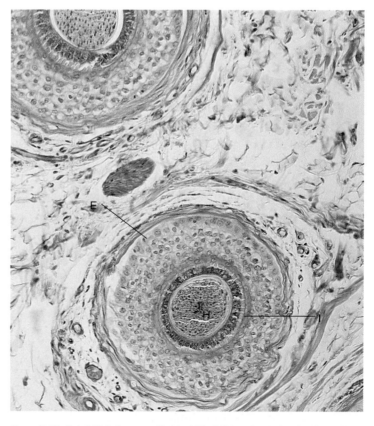

Figure 5-42 *Hair follicle (cross section)* (×132). *E,* External root sheath; *I,* internal root sheath; *H,* hair.

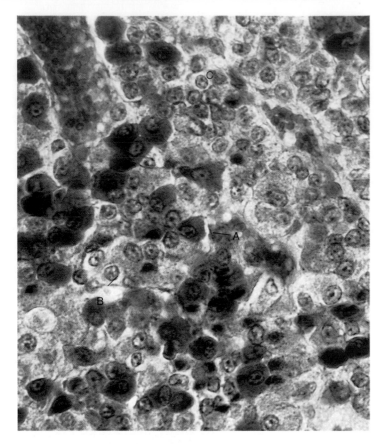

Figure 5-43 *Pituitary gland (anterior)* (×470). *A,* Acidophil; *B,* basophil; *C,* chromophil.

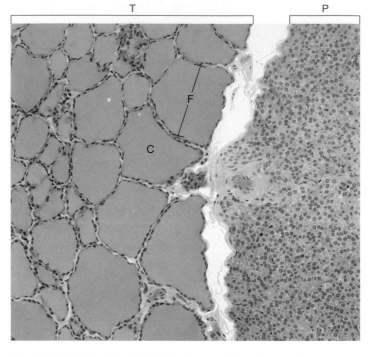

Figure 5-44 *Thyroid and parathyroid gland* (×132). *C,* Thyroid colloid; *F,* thyroid follicle; *P,* parathyroid gland; *T,* thyroid gland.

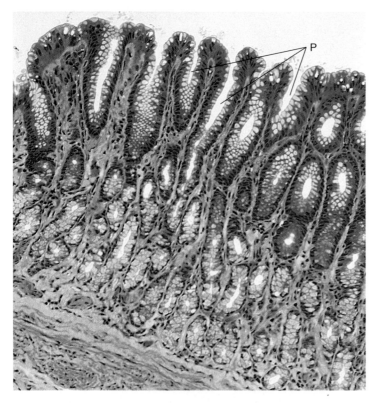

Figure 5-45 *Stomach lining (pylorus)* (×132). *P,* Gastric pits.

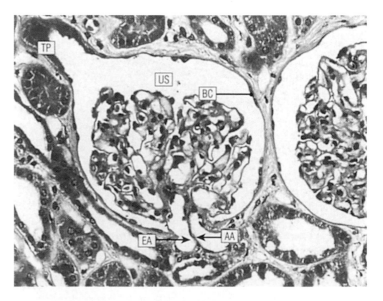

Figure 5-46 *Renal corpuscle of kidney. AA,* Afferent arteriole; *EA,* efferent arteriole; *TP,* proximal tubule; *US,* urinary space of Bowman's capsule; *BC,* parietal wall of Bowman's capsule.

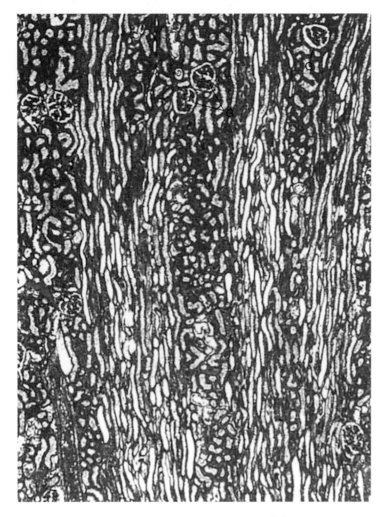

Figure 5-47 *Kidney tubules and blood vessels.* G, Glomeruli.

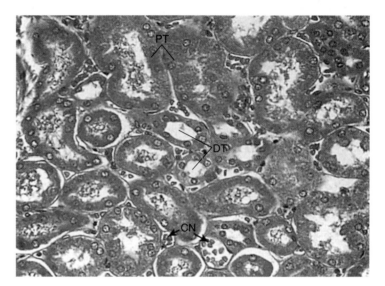

Figure 5-48 *Kidney tubules (cross section).* PT, Proximal tubules; DT, distal tubules; CN, capillary networks.

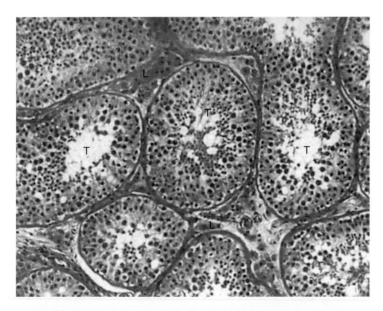

Figure 5-49 *Seminiferous tubules of testis (cross section).* T, Seminiferous tubules; L, interstitial (Leydig) cells.

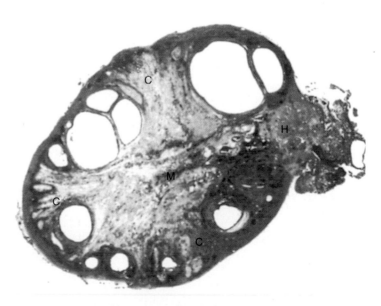

Figure 5-50 *Ovary (cross section).* H, Hilum; C, cortex; M, medulla.

Photograph/Illustration Credits

BRIEF ATLAS: PART 1

Figures 1-1 to 1-36: Drake RL et al: *Gray's atlas of anatomy*, ed 2, Philadelphia, 2015, Elsevier.

BRIEF ATLAS: PART 2

Figures 2-1 to 2-11: Abrahams PH: *McMinn's color atlas of human anatomy*, ed 5, Philadelphia, 2003, Elsevier.

Figures 2-12 to 2-91: Muscolino JE: *Kinesiology: the skeletal system and muscle function*, ed 2, St Louis, 2010, Elsevier.

BRIEF ATLAS: PART 3

Figures 3-1 to 3-22: Abrahams PH: *McMinn's color atlas of human anatomy*, ed 5, Philadelphia, 2003, Elsevier.

BRIEF ATLAS: PART 4

Figures 4-1 to 4-8: Abrahams PH: *McMinn's color atlas of human anatomy*, ed 5, Philadelphia, 2003, Elsevier.

Figures 4-9 to 4-16: Gosling JA: *Human anatomy color atlas and text*, ed 4, London, 2002, Elsevier.

Figures 4-17 to 4-24: Moses K: *Atlas of clinical gross anatomy*, Edinburgh, 2005, Elsevier.

BRIEF ATLAS: PART 5

Figure 5-1: Courtesy Don Fawcett, Harvard Medical School.

Figure 5-2: Pollard TD, Earnshaw WC: *Cell biology*, ed 2, Philadelphia, 2007, Elsevier.

Figures 5-3, 5-4, 5-5, 5-6, 5-7, 5-8, 5-9, 5-10, 5-11, 5-12, 5-13, 5-19, 5-20, 5-21, 5-22, 5-23, 5-25, 5-26, 5-27, 5-28, 5-29, 5-30, 5-31, 5-32, 5-34, 5-35, 5-36, 5-37, 5-39, 5-40, 5-41, 5-42, 5-43, 5-44, 5-45: Gartner LP, Hiatt JL: *Color textbook of histology*, ed 2, Burlington, 2001, Elsevier.

Figures 5-14, 5-15, 5-24, 5-38: Leeson TS, Leeson CR, Paparo AA: *Text/atlas of histology*, Philadelphia, 1988, Elsevier.

Figures 5-16, 5-17, 5-18, 5-33, 5-46, 5-47, 5-48, 5-49, 5-50: Stevens A, Lowe JS: *Human histology*, ed 3, Philadelphia, 2005, Elsevier.

Quick Guide to the Language of Science and Medicine

If you are unfamiliar with it, scientific and medical terminology can seem overwhelming. The length and apparent complexity of many terms often scare people who have not had any training or practice in scientific terminology. Although it requires knowledge of some basic word parts and a few rules for using them, scientific terminology is not as difficult as it seems. This handy, removable insert provides what you need to get you started.

First, there are a handful of hints to help you learn and use the terms we will use in *Anatomy & Physiology*. Second, there are several tables containing many of the most commonly used word parts and examples of how they are used. This handy summary does not attempt to teach you the entire field of scientific and medical terminology, but with the information given here and a little practice, you will soon become comfortable with the basics.

LEARNING AND USING SCIENTIFIC TERMS

MOST TERMS COME FROM LATIN AND GREEK

Many scientific terms are derived from the Latin and Greek languages. This is because many of the anatomists, physiologists, and physicians who discovered the basic principles of modern life science used these languages themselves so that they could communicate with each other without having to learn dozens of individual native languages. Thus Latin and Greek have become the "universal" language of scientific terminology. Many of the words used are derived from these classical languages, as are some of their rules of use. The most useful rules are given in a later section.

TERMS ARE MADE BY COMBINING WORD PARTS

One set of rules for using Latin and Greek is essential to understanding scientific terminology. Both languages rely on the ability to combine word parts to make new words. Thus almost all scientific terms are constructed by combining smaller word elements to make a specific term. Because of this combining technique, many scientific terms appear at first glance to be long and complex. However, if you read a new term as a series of word elements rather than a single word, determining the meaning will be less imposing. One of the easiest ways to learn scientific terminology is to develop the ability to analyze new terms instantly to discover the word parts that comprise them. Different kinds of word parts are used depending on exactly how they fit with other word parts to form a complete term.

- A **prefix** is a word part added to the beginning of an existing word to alter its meaning. We use prefixes in English as well; for example, the meaning of *sense* changes when the prefix *non-* is added to make the word *nonsense*.
- A **suffix** is a word part added to the end of an existing word to alter its meaning. Once again, suffixes are also sometimes used in English. For example, the meaning of *sense* changes when the suffix *-less* is added to make the word *senseless*. A complex term can have a series of suffixes, a series of prefixes, or both. For example, the word *senselessness* has two suffixes: *-less* and *-ness*.
- A **root** is a word part that serves as the starting point for forming a new term. In the previous examples in English, the word *sense* was the root to which was added a prefix or a suffix. Word parts commonly used as roots can also be used as suffixes or prefixes in forming a new term. Also, several roots are sometimes combined to form a larger root to which suffixes or prefixes can be added.
- **Combining vowels** are vowels (*a, e, i, o, u, y*) used to link word parts, often to make pronunciation easier. For example, to link the suffix *-tion* to the root *sense*, we must use the combining vowel *-a-* to form the new term *sensation*. Using the *-e* that is already there would make the term difficult to pronounce. A root and a combining vowel together, such as *sensa-*, are often called the **combining form** of the word part.

As in English, Latin and Greek word parts have synonyms. Therefore a word part may have different meanings in different contexts, and different word parts may all have a similar meaning. This adds to the richness of scientific terminology, as it does in any language.

 Hint In both the chapter word lists (Language of Science and Language of Medicine) and glossary of the *Anatomy & Physiology* textbook, we have listed the word parts and their meanings for each entry. If you read through these word parts as you encounter these entries, you will begin to learn them quickly and easily.

SOME TERMS USE LATIN PLURAL FORMS

Another set of rules for using Latin-based and Greek-based terms that you will find useful pertains to pluralization. To form a plural in English, we often simply add *-s* or *-es* to a word. For example, the plural for *sense* is *senses*. Because we have adopted scientific terms into English to form sentences, we often simply use the pluralization rules of English and add the *-s* or *-es*. At times, however, you will run across an English term that has been pluralized according to Latin or Greek rules. As in ordinary English, some Latin-based forms are the same whether they are used as singular or plural. For example, the term *meatus* (a tubelike opening) is both singular and plural. This brief list will help you distinguish between many plural and singular forms:

Singular	Plural	Example
-a	-ae	Ampulla, ampullae
-ax	-aces	Thorax, thoraces
-en	-ina	Foramen, foramina
-ex	-ices	Cortex, cortices
-is	-es	Neurosis, neuroses
-ix	-ices	Appendix, appendices
-ma	-mata	Lymphoma, lymphomata
-on	-a	Mitochondrion, mitochondria
-um	-a	Datum, data
-ur	-ora	Femur, femora
-us	-i	Villus, villi
-yx	-yces	Calyx, calyces

AVOID CONFUSING ADJECTIVES WITH NOUNS

As in any communication in English, we sometimes convert nouns to adjectives to describe anatomical parts or physiological processes. This is usually done by adding the word parts *-ic* or *-al* to the end of a noun to make it an adjective. For example, if we want to describe something pertaining to the **base** of a structure, we could use **bas<u>al</u>** or **bas<u>ic</u>**—as in *basal layer* or *basic solution*. Sometimes we even go crazy and add both endings, as in *anatomi<u>c</u>al* (the adjective form of the noun *anatomy*).

It would be accurate to state that the *femur* is a bone in the *femoral* region of the lower extremity or that the *facial nerve* has connections in the *face*.

In using these terms, the noun form can stand alone but the adjective form needs a noun to modify (describe). For example, it is not proper to state "carpal" when you mean "carpal bones" or "carpus." Although stating "My carpals hurt" is often used in casual communictation, stating "My carpal bones hurt" is more accurate and less likely to be misinterpreted. And if you really mean the whole wrist, it would be best to state "My carpal region hurts" or "My carpus hurts."

Likewise, referring to your "occipital" is not as clear as referring to your "occipital bone" or "occiput."

CORRECT SPELLING IS IMPORTANT

Correct spelling of scientific terms is essential to their meanings. This is especially true of terms that are very close in spelling but very different in meaning. For example, the *perineum* is the region of the trunk around the genitals and anus, whereas the *peritoneum* is a membrane that lines the abdominal cavity and covers abdominal organs. Even a mistake in one letter can change the meaning of a word, as in the case of *ilium* (part of the bony pelvis) and *ileum* (part of the small intestine).

American spellings of terms occasionally differ from other forms of English, such as in Canada, Australia, or the United Kingdom. For example *centimeter*, the American form, and *centimetre* are both correct. The American *esophagus* is often spelled *oesophagus* in other parts of the world. Always use the spelling appropriate for your location and intended audience. Here are some examples of alternate spellings that may be used in anatomy and physiology discussions:

American (U.S.)	United Kingdom
Aging	Ageing
Anemia, anemic	Anaemia, anaemic
Anesthetic	Aenesthetic
Cecum	Caecum
Celiac	Coeliac
Center	Centre
Color	Colour
Diarrhea	Diarrhoea
Edema	Edaema
Esophagus	Oesophagus
Estrogen	Oestrogen
Fecal	Faecal
Fetal, fetus	Foetal, foetus
Fiber	Fibre
Filter	Filtre
Flavor	Flavour
Gray	Grey
Hemoglobin, hemo-	Haemoglobin, haemo-
Humor	Humour
Ischemia, ischemic	Ischaemia, ischaemic
Labor	Labour
Leukemia	Leukaemia
Liter, -liter	Litre, -litre
Meter, -meter	Metre, -metre
Mold	Mould
Neuron	Neurone
Pediatric	Paediatric
Sensitize	Sensitise
Sulfate, sulfur	Sulphate, sulphur
Tenia	Taenia
Tumor	Tumour

CORRECT PRONUNCIATION IS IMPORTANT

Because many scientific terms are spoken, correct pronunciation is as important as correct spelling. Scientific terms can usually be pronounced phonetically—by sounding out each letter sound of each syllable. Check the pronunciation keys given in each chapter and in the glossary if you are uncertain of how to pronounce any word presented in *Anatomy & Physiology*. Audio pronunciations are also provided with the electronic resources that accompany this textbook.

Regional differences in pronunciation and differences among different branches of science and medicine do exist. For example, the American pronunciation of vitamin is *VYE-tah-min*, but in many other parts of the world the pronunciation is *VIH-tah-min*. The term *centimeter* is correctly pronounced *SEN-tah-mee-ter*, but some health professionals have learned the pronunciation *SAWN-tah-mee-ter*.

BE AWARE OF ALTERNATE TERMINOLOGY

As explained in Chapter 1 of the text, the *Terminologia Anatomica* (Ta or TA) is an international list of gross anatomy terms and the *Terminologia Histologica* (Th or TH) is an international list of microscopic anatomy terms. Other international lists are also in development, such as the *Terminologia Embryologica* (Te or TE) for developmental anatomy terms. Although such lists are helpful, alternate terminology is frequently encountered. For example, the standard lists themselves often show alternate forms of the same term. Good reasons sometimes exist for using a newer term or a term that is more easily understood by your colleagues or in your region of the world. Do not forget that new terminology is slow to be adopted worldwide, so you may need to know newer and older terms to communicate easily with many different people.

Also remember that eponyms, or terms based on the name of a person, are also frequently used in science and medicine. However, most eponyms are currently being phased out in favor of more descriptive terms. A brief list of some important eponyms and their alternatives is given later in this section.

When using eponyms, it is an increasingly common practice to drop the apostrophe and "s" at the end of an eponym. In other words, we do not use the possessive form of the name. For example, we use *Bowman capsule* instead of *Bowman's capsule*, or *Parkinson disease* rather than *Parkinson's disease*. Likewise, *organ of Corti* is increasingly replaced by *Corti organ*. But be aware that not everyone follows this practice, so you will sometimes encounter the possessive form.

PRACTICE NEW TERMINOLOGY

As you know, practice makes perfect. Practice using the scientific terms in this or another book until you become comfortable with scientific terminology. It won't take long, and you will probably have fun doing it.

 One of the easiest ways to practice new terms quickly and painlessly is to use flash cards. Write each term on a separate index card or slip of paper. On the other side of each card, write what you need to know about the term. By doing this step you will have already learned quite a bit! Carry the deck of cards you made with you everywhere. When you get a few minutes here and there, take out the deck and look at the term on top. Try to guess its meaning. Check yourself by flipping the card over. Do not fret if you get it wrong, just bury the card in the deck and move on. After a while, you will find it "sinking in" and you will get them all correct. There are also many web-based flash-card apps for computers and smart phones available. Go to theAPstudent.org for more hints on practicing terminology.

TABLE 1   **Word Parts Commonly Used as Prefixes**

 Prefixes are word parts that are added *before* the root. Prefixes may be "stacked" with other prefixes to produce the final meaning of a term. The meanings given here are common meanings; in different contexts, the meaning may be a bit different than listed here.

WORD PART	MEANING	EXAMPLE	MEANING OF EXAMPLE
a-	Without, not	Apnea	Cessation of breathing
a[d]-	Toward	Afferent	Carrying toward
all[o]-	[an]other, different	Allosteric	Another shape
an-	Without, not	Anuria	Absence of urination
ante-	Before	Antenatal	Before birth
anti-	Against, resisting	Antibody	Unit that resists foreign substances
auto-	Self	Autoimmunity	Self-immunity
bi-	Two, double	Bicuspid	Two-pointed
circum-	Around	Circumcision	Cutting around
co-, con-	With, together	Congenital	Born with
contra-	Against	Contraceptive	Against conception
de-	Down from, undoing	Defibrillation	Stop fibrillation
dia-	Across, through	Diarrhea	Flow through (intestines)
dipl-	Twofold, double	Diploid	Two sets of chromosomes
dys-	Bad, disordered, difficult	Dysplasia	Disordered growth
ectop-	Displaced	Ectopic pregnancy	Displaced pregnancy
ef-	Away from	Efferent	Carrying away from
em-, en-	In, into	Encyst	Enclose in a cyst
endo-	Within	Endocarditis	Inflammation of heart lining
epi-	Upon	Epimysium	Covering of a muscle
eu-	Good	Eupnea	Good (normal) breathing
ex-, exo-	Out of, out from	Exophthalmos	Protruding eyes
extra-	Outside of	Extraperitoneal	Outside the peritoneum
hapl-	Single	Haploid	Single set of chromosomes

Continued

T A B L E   1   Word Parts Commonly Used as Prefixes—cont'd

WORD PART	MEANING	EXAMPLE	MEANING OF EXAMPLE
hem-, hemat-	Blood	Hematuria	Bloody urine
hemi-	Half	Hemiplegia	Paralysis in half the body
hom(e)o-	Same, equal	Homeostasis	Standing the same
hyper-	Over, above	Hyperplasia	Excessive growth
hypo-	Under, below	Hypodermic	Below the skin
infra-	Below, beneath	Infraorbital	Below the (eye) orbit
inter-	Between	Intervertebral	Between vertebrae
intra-	Within	Intracranial	Within the skull
iso-	Same, equal	Isometric	Same length
macro-	Large	Macrophage	Large eater (phagocyte)
mega-	Large, million(th)	Megakaryocyte	Cell with large nucleus
mes-	Middle	Mesentery	Middle of intestine
meta-	Beyond, after, change, middle	Metatarsal	Beyond the tarsals (ankle bones)
micro-	Small, millionth	Microcytic	Small-celled
milli-	Thousandth	Milliliter	Thousandth of a liter
mono-	One (single)	Monosomy	Single chromosome
neo-	New	Neoplasm	New matter
non-	Not	Nondisjunction	Not disjoined
oligo-	Few, scanty	Oliguria	Scanty urination
ortho-	Straight, correct, normal	Orthopnea	Normal breathing
para-	By the side of, near	Parathyroid	Near the thyroid
per-	Through	Permeable	Able to go through
peri-	Around, surrounding	Pericardium	Covering of the heart
poly-	Many	Polycythemia	Condition of many blood cells
post-	After	Postmortem	After death
pre-	Before	Premenstrual	Before menstruation
pro-	First, promoting	Progesterone	Hormone that promotes pregnancy
quadr-	Four	Quadriplegia	Paralysis in four limbs
re-	Back again	Reflux	Backflow
retro-	Behind	Retroperitoneal	Behind the peritoneum
semi-	Half	Semilunar	Half-moon
sub-	Under	Subcutaneous	Under the skin
super-, supra-	Over, above, excessive	Superior	Above
trans-	Across, through	Transcutaneous	Through the skin
tri-	Three, triple	Triplegia	Paralysis of three limbs

TABLE 2   Word Parts Commonly Used as Suffixes

> **Hint** Suffixes are word parts that are added *after* the root. Prefixes may be "stacked" with other suffixes to produce the final meaning of a term. The meanings given here are common meanings; in different contexts, the meaning may be a bit different than listed here.

WORD PART	MEANING	EXAMPLE	MEANING OF EXAMPLE
-al, -ac	Pertaining to	Intestinal	Pertaining to the intestines
-algia	Pain	Neuralgia	Nerve pain
-aps, -apt	Fit, fasten	Synapse	Fasten together
-arche	Beginning, origin	Menarche	First menstruation
-ase	Signifies an enzyme	Lipase	Enzyme that acts on lipids
-blast	Sprout, make	Osteoblast	Bone maker
-centesis	A piercing	Amniocentesis	Piercing the amniotic sac
-cide	To kill	Fungicide	Fungus killer
-clast	Break, destroy	Osteoclast	Bone breaker
-crine	Release, secrete	Endocrine	Secrete within
-ectomy	A cutting out	Appendectomy	Removal of the appendix
-emesis	Vomiting	Hematemesis	Vomiting blood
-emia	Refers to blood condition	Hypercholesterolemia	High blood cholesterol level
-flux	Flow	Reflux	Backflow
-gen	Creates, forms	Lactogen	Milk producer
-genesis	Origin, production	Oogenesis	Egg production
-gram*	Something written	Electroencephalogram	Record of brain's electrical activity
-graph(y)*	To write, draw	Electrocardiograph	Apparatus that records heart's electrical activity
-hydrate	Containing H_2O (water)	Dehydration	Loss of water
-ia, -sia	Condition, process	Arthralgia	Condition of joint pain
-iasis	Abnormal condition	Giardiasis	*Giardia* infestation
-ic, -ac	Pertaining to	Cardiac	Pertaining to the heart
-in	Signifies a protein	Renin	Kidney protein
-ism	Signifies "condition of"	Gigantism	Condition of gigantic size
-itis	Signifies "inflammation of"	Gastritis	Stomach inflammation
-lemma	Rind, peel	Neurilemma	Covering of a nerve fiber
-lepsy	Seizure	Epilepsy	Seizure upon seizure
-lith	Stone, rock	Lithotripsy	Stone-crushing

*A term ending in *-graph* refers to an apparatus that results in a visual and/or recorded representation of biological phenomena, whereas a term ending in *-graphy* is the technique or process of using the apparatus. A term ending in *-gram* is the record itself. For example, in electrocardio*graphy,* an electrocardio*graph* is used in producing an electrocardio*gram.*   *Continued*

TABLE 2   Word Parts Commonly Used as Suffixes—cont'd

WORD PART	MEANING	EXAMPLE	MEANING OF EXAMPLE
-logy	Study of	Cardiology	Study of the heart
-lunar	Moon, moonlike	Semilunar	Half-moon
-malacia	Softening	Osteomalacia	Bone softening
-megaly	Enlargement	Splenomegaly	Spleen enlargement
-metric, -metry	Measurement, length	Isometric	Same length
-oid	Like, in the shape of	Sigmoid	S-shaped
-oma	Tumor	Lipoma	Fatty tumor
-opia	Vision, vision condition	Myopia	Nearsightedness
-oscopy	Viewing	Laparoscopy	Viewing the abdominal cavity
-ose	Signifies a carbohydrate (especially sugar)	Lactose	Milk sugar
-osis	Condition, process	Dermatosis	Skin condition
-ostomy	Formation of an opening	Tracheostomy	Forming an opening in the trachea
-otomy	Cut	Lobotomy	Cut of a lobe
-penia	Lack	Leukopenia	Lack of white (cells)
-philic	Loving	Hydrophilic	Water-loving
-phobic	Fearing	Hydrophobic	Water-fearing
-phragm	Partition	Diaphragm	Partition separating thoracic and abdominal cavities
-plasia	Growth, formation	Hyperplasia	Excessive growth
-plasm	Substance, matter	Neoplasm	New matter
-plasty	Shape, make	Rhinoplasty	Reshaping the nose
-plegia	Paralysis	Triplegia	Paralysis in three limbs
-pnea	Breath, breathing	Apnea	Cessation of breathing
-(r)rhage, -(r)rhagia	Breaking out, discharge	Hemorrhage	Blood discharge
-(r)rhaphy	Sew, suture	Meningeorrhaphy	Suturing of meninges
-(r)rhea	Flow	Diarrhea	Flow through (intestines)
-some	Body	Chromosome	Stained body
-tensin, -tension	Pressure	Hypertension	High pressure
-tonic	Relating to pressure, tension	Isotonic	Same pressure
-tripsy	Crushing	Lithotripsy	Stone-crushing
-ule	Small, little	Tubule	Small tube
-uria	Refers to urine condition	Proteinuria	Protein in the urine

TABLE 3   Word Parts Commonly Used as Roots

Hint ▶ Roots are word parts that are the basis of a term; they often have a prefix or suffix added to them. Roots may be combined with other roots, or used as prefixes or suffixes, to produce the final meaning of a term. The meanings given here are common meanings; in different contexts, the meaning may be a bit different than listed here.

WORD PART	MEANING	EXAMPLE	MEANING OF EXAMPLE
-acro-	Extremity	Acromegaly	Enlargement of extremities
-aden-	Gland	Adenoma	Tumor of glandular tissue
-alveol-	Small hollow, cavity	Alveolus	Small air sac in the lung
-angi-	Vessel	Angioplasty	Reshaping a vessel
-arthr-	Joint	Arthritis	Joint inflammation
-asthen-	Weakness	Myasthenia	Condition of muscle weakness
-bar-	Pressure	Baroreceptor	Pressure receptor
-bili-	Bile	Bilirubin	Orange-yellow bile pigment
-brachi-	Arm	Brachial	Pertaining to the arm
-brady-	Slow	Bradycardia	Slow heart rate
-bronch-	Air passage	Bronchitis	Inflammation of pulmonary passages (bronchi)
-calc-	Calcium, limestone	Hypocalcemia	Low blood calcium level
-capn-	Smoke	Hypercapnia	Elevated blood CO_2 level
-carcin-	Cancer	Carcinogen	Cancer producer
-card-	Heart	Cardiology	Study of the heart
-cephal-	Head, brain	Encephalitis	Brain inflammation
-cerv-	Neck	Cervicitis	Inflammation of (uterine) cervix
-chem-	Chemical	Chemotherapy	Chemical treatment
-chol-	Bile	Cholecystectomy	Removal of bile (gall)bladder
-chondr-	Cartilage	Chondroma	Tumor of cartilage tissue
-chrom-	Color	Chromosome	Stained body
-corp-	Body	Corpus luteum	Yellow body
-cortico-	Pertaining to cortex	Corticosteroid	Steroid secreted by (adrenal) cortex
-crani-	Skull	Intracranial	Within the skull
-crypt-	Hidden	Cryptorchidism	Undescended testis
-cusp-	Point	Tricuspid	Three-pointed
-cut(an)-	Skin	Transcutaneous	Through the skin
-cyan-	Blue	Cyanosis	Condition of blueness
-cyst-	Bladder	Cystitis	Bladder inflammation
-cyt-	Cell	Cytotoxin	Cell poison
-dactyl-	Fingers, toes (digits)	Syndactyly	Joined digits
-dendr-	Tree, branched	Oligodendrocyte	Branched nervous tissue cell
-dent-	Tooth	Dentalgia	Toothache
-derm-	Skin	Dermatitis	Skin inflammation
-diastol-	Relax, stand apart	Diastole	Relaxation phase of heartbeat
-dips-	Thirst	Polydipsia	Excessive thirst
-ejacul-	To throw out	Ejaculation	Expulsion (of semen)
-electr-	Electrical	Electrocardiogram	Record of electrical activity of heart
-enter-	Intestine	Enteritis	Intestinal inflammation
-eryth(r)-	Red	Erythrocyte	Red (blood) cell

Continued

TABLE 3   **Word Parts Commonly Used as Roots—cont'd**

WORD PART	MEANING	EXAMPLE	MEANING OF EXAMPLE
-esthe-	Sensation	Anesthesia	Condition of no sensation
-febr-	Fever	Febrile	Pertaining to fever
-gastr-	Stomach	Gastritis	Stomach inflammation
-gest-	To bear, carry	Gestation	Pregnancy
-gingiv-	Gums	Gingivitis	Gum inflammation
-glomer-	Wound into a ball	Glomerulus	Rounded tuft of vessels
-gloss-	Tongue	Hypoglossal	Under the tongue
-gluc-	Glucose, sugar	Glucosuria	Glucose in urine
-glutin-	Glue	Agglutination	Sticking together (of particles)
-glyc-	Sugar (carbohydrate), glucose	Glycolipid	Carbohydrate-lipid combination
-hepat-	Liver	Hepatitis	Liver inflammation
-hist-	Tissue	Histology	Study of tissues
-hydro-	Water	Hydrocephalus	Water on the brain
-hyster-	Uterus	Hysterectomy	Removal of the uterus
-iatr-	Treatment	Podiatry	Foot treatment
-kal-	Potassium	Hyperkalemia	Elevated blood potassium level
-kary-	Nucleus	Karyotype	Array of chromosomes from nucleus
-kerat-	Cornea	Keratotomy	Cutting of the cornea
-kin-	To move, divide	Kinesthesia	Sensation of body movement
-lact-	Milk, milk production	Lactose	Milk sugar
-lapar-	Abdomen	Laparoscopy	Viewing the abdominal cavity
-leuk-	White	Leukorrhea	White flow (discharge)
-lig-	To tie, bind	Ligament	Tissue that binds bones
-lip-	Lipid (fat)	Lipoma	Fatty tumor
-lys-	Break apart	Hemolysis	Breaking of blood cells
-mal-	Bad	Malabsorption	Improper absorption
-melan-	Black	Melanin	Black protein
-men-, -mens-, (-menstru-)	Month (monthly)	Amenorrhea	Absence of monthly flow
-metr-	Uterus	Endometrium	Uterine lining
-muta-	Change	Mutagen	Change maker
-my-, -myo-	Muscle	Myopathy	Muscle disease
-myc-	Fungus	Mycosis	Fungal condition
-myel-	Marrow	Myeloma	(Bone) marrow tumor
-myx-	Mucus	Myxedema	Mucous edema
-nat-	Birth	Neonatal	Pertaining to newborns (infants)
-natr-	Sodium	Natriuresis	Elevated sodium in urine
-nephr-	Nephron, kidney	Nephritis	Kidney inflammation
-neur-	Nerve	Neuralgia	Nerve pain
-noct-, -nyct-	Night	Nocturia	Urination at night
-ocul-	Eye	Binocular	Two-eyed
-odont-	Tooth	Periodontitis	Inflammation (of tissue) around the teeth
-onco-	Cancer	Oncogene	Cancer gene

T A B L E 3   Word Parts Commonly Used as Roots—cont'd

WORD PART	MEANING	EXAMPLE	MEANING OF EXAMPLE
-ophthalm-	Eye	Ophthalmology	Study of the eye
-orchid-	Testis	Orchiditis	Testis inflammation
-osteo-	Bone	Osteoma	Bone tumor
-oto-	Ear	Otosclerosis	Hardening of ear tissue
-ov-, -oo-	Egg	Oogenesis	Egg production
-oxy-	Oxygen	Oxyhemoglobin	Oxygen-hemoglobin combination
-path-	Disease	Neuropathy	Nerve disease
-ped-	Children	Pediatric	Pertaining to treatment of children
-phag-	Eat	Phagocytosis	Cell eating
-pharm-	Drug	Pharmacology	Study of drugs
-phleb-	Vein	Phlebitis	Vein inflammation
-photo-	Light	Photopigment	Light-sensitive pigment
-physio-	Nature (function) of	Physiology	Study of biological function
-pino-	Drink	Pinocytosis	Cell drinking
-plex-	Twisted, woven	Nerve plexus	Complex of interwoven nerve fibers
-pneumo-	Air, breath	Pneumothorax	Air in the thorax
-pneumon-	Lung	Pneumonia	Lung condition
-pod-	Foot	Podocyte	Cell with feet
-poie-	Make, produce	Hemopoiesis	Blood cell production
-pol-	Axis, having poles	Bipolar	Having two ends
-presby-	Old	Presbyopia	Old vision
-proct-	Rectum	Proctoscope	Instrument for viewing the rectum
-pseud-	False	Pseudopodia	False feet
-psych-	Mind	Psychiatry	Treatment of the mind
-pyel-	Pelvis	Pyelogram	Image of the renal pelvis
-pyo-	Pus	Pyogenic	Pus-producing
-pyro-	Heat, fever	Pyrogen	Fever producer
-ren-	Kidney	Renocortical	Referring to the cortex of the kidney
-rhino-	Nose	Rhinoplasty	Reshaping the nose
-rigor-	Stiffness	Rigor mortis	Stiffness of death
-sarco-	Flesh, muscle	Sarcolemma	Muscle fiber membrane
-scler-	Hard	Scleroderma	Hard skin
-semen-, -semin-	Seed, sperm	Seminiferous tubule	Sperm-bearing tubule
-sept-	Contamination	Septicemia	Contamination of the blood
-sigm-	Σ or Roman S	Sigmoid colon	S-shaped colon
-sin-	Cavity, recess	Paranasal sinus	Cavity near the nasal cavity
-son-	Sound	Sonography	Imaging using sound
-spiro-, -spire	Breathe	Spirometry	Measurement of breathing
-stat-, -stas-	A standing, stopping	Homeostasis	Staying the same
-syn-	Together	Syndrome	Signs appearing together
-systol-	Contract, stand together	Systole	Contraction phase of the heartbeat
-tachy-	Fast	Tachycardia	Rapid heart rate

Continued

TABLE 3 **Word Parts Commonly Used as Roots—cont'd**

WORD PART	MEANING	EXAMPLE	MEANING OF EXAMPLE
-therm-	Heat	Thermoreceptor	Heat receptor
-thromb-	Clot	Thrombosis	Condition of abnormal blood clotting
-tom-	A cut, a slice	Tomography	Image of a slice or section
-tox-	Poison	Cytotoxin	Cell poison
-troph-	Grow, nourish	Hypertrophy	Excessive growth
-tympan-	Drum	Tympanum	Eardrum
-varic-	Enlarged vessel	Varicose vein	Enlarged vein
-vas-	Vessel, duct	Vasoconstriction	Vessel narrowing
-vesic-	Bladder, blister	Vesicle	Blister
-vol-	Volume	Hypovolemic	Characterized by low volume

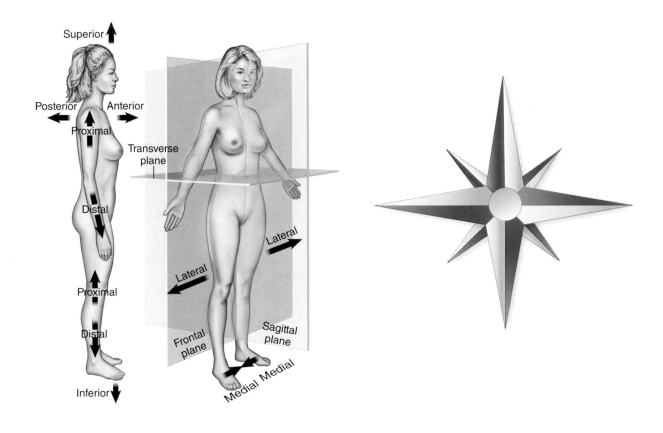

TABLE 4 Abbreviations Used for Anatomical Directions

 To minimize confusion when discussing the relations between body areas or the location of a particular anatomical structure, specific directional terms are used. To make the reading of anatomical figures a little easier, an *anatomical rosette* is used in *Anatomy & Physiology*. On many figures you will notice a small compass rosette similar to those on geographical maps. Rather than being labeled N, S, E, and W, the anatomical rosette is labeled with abbreviated anatomical directions as demonstrated in the figures on p. 10 and the table below.

ABBREVIATION	DIRECTION*	DESCRIPTION	EXAMPLES OF USAGE
A	Anterior	Front; in front of	The navel is anterior to the abdominal cavity. The nose is on the anterior surface of the head. The anterior portion of the brain's cerebrum is called the frontal lobe.
P (opposite A)	Posterior	Back; in back of	The buttocks is posterior to the genitals. The heart is posterior to the sternum (breastbone). The posterior portion of the brain's cerebrum is called the occipital lobe.
P (opposite D)	Proximal	Toward the origin	The shoulder is proximal to the elbow. The knee is proximal to the ankle. The proximal tubule of the kidney is near the beginning of the tubule.
D	Distal	Away from the origin	The foot is distal to the tibia. The wrist is distal to the humerus. The distal tubule of the kidney is near the end of the tubule
S	Superior	Upper; above	The lungs are superior to the diaphragm. The head is at the superior end of the body. The atria are the superior chambers of the heart.
I	Inferior	Lower; below	The stomach is inferior to the diaphragm. The pelvic cavity is inferior to the abdominal cavity. The ventricles are the inferior chambers of the heart.
M	Medial	Toward the midline of the body or organ	The tibia is medial to the fibula in the leg. A medial indentation is found on each kidney. The medial condyle is on the side of the bone toward the middle of the body.
L (opposite M)	Lateral	Away from the midline of the body or organ; toward the side	The radius is lateral to the ulna in the anatomical position. The lateral ventricles are fluid chambers on either side of the midline of the brain. A lateral process of a vertebra lies on the extreme left and extreme right of the bone.
L (opposite R)	Left	On or toward the left side	The left ventricle of the heart has a thicker wall than the right ventricle. The left eardrum is infected. The lungs lie to the right and left of the heart.
R	Right	On or toward the right side	The right lung is usually larger than the left lung. The liver is located on the right side of the abdomen. The right femoral artery is damaged.

*See the inside front cover of *Anatomy & Physiology* for additional directional terms.

 These terms relate to the body as if it is in the anatomical position (upright, facing forward, hands to the side and palms forward, feet forward and slightly apart). Be especially careful with the terms left and right. They refer to a *subject's* left or right—not *your* left or right.

TABLE 5   **Common Eponyms and Equivalent Descriptive Terms**

 Eponyms are terms named for people or places. Current practice is to reduce or eliminate the use of eponyms in favor of descriptive terms. For example, *cells of Leydig* are now more often called *interstitial cells*. If the eponym is used, however, the possessive or genitive form of the eponym is typically dropped. Thus instead of *cells of Leydig*, we would use *Leydig cells*. This also applies to diseases and disorders, so *Alzheimer's disease* is now *Alzheimer disease*.

EPONYM	DESCRIPTIVE TERM	LOCATION	EPONYM	DESCRIPTIVE TERM	LOCATION
Achilles tendon	Calcaneal tendon	Ankle and foot	Müllerian duct	Paramesonephric duct	Urogenital region (embryo)
Adam's apple	Thyroid cartilage	Larynx	Nissl substance	Chromatophilic substance	Neuron
Auerbach plexus	Myenteric plexus	Intestinal wall	Oddi sphincter	Sphincter of hepatopancreatic ampulla	Common bile duct
Bartholin gland	Greater vestibular gland	Vulva			
Bowman capsule	Glomerular capsule	Kidney	Pacinian corpuscle	Lamellar corpuscle	Skin
Broca area	Motor speech area	Brain	Peyer patch	(Regional) lymphoid nodules	Intestinal lining
Brunner gland	Duodenal gland	Small intestine			
Cooper ligament	Suspensory ligament	Breast	Purkinje fiber	Subendocardial branch	Heart
Cori cycle	Lactic acid cycle, lactate cycle	Muscle and liver cells	Ranvier node	Neurofibral node	Nerve tissue
Corti organ	Spiral organ	Inner ear	Reissner membrane	Vestibular membrane	Inner ear
Cowper gland	Bulbourethral gland	Male urethra	Rivinus duct	Sublingual duct	Mouth
Eustachian tube	Auditory tube	Ear and pharynx	Rolando fissure	Central sulcus	Brain
Fallopian tube	Uterine tube	Female reproductive tract	Rotter node	Interpectoral lymph node	Thoracic wall
Golgi body (complex)	Dictyosome	Cell	Ruffini corpuscle	Bulbous corpuscle	Skin
Graaf(ian) follicle	Vesicular ovarian follicle	Ovary	Santorini duct	Accessory pancreatic duct	Pancreas
Hassall corpuscle	Thymic corpuscle	Thymus	Schlemm canal	Scleral venous sinus	Eye
Havers(ian) canal	Central canal	Bone tissue	Schwann cell	Neurilemmocyte	Nerve tissue
Havers(ian) system	Osteon	Bone tissue	Sertoli cell	Sustentacular cells; nurse cell	Testis
Henle loop	Nephron loop	Kidney			
His bundle	Atrioventricular (AV) bundle	Heart wall	Sharpey fiber	Perforating collagen fiber bundle	Bone tissue
Krebs cycle	Citric acid cycle, tricarboxylic acid (TCA) cycle	Mitochondrion	Skene gland	Paraurethral gland, female prostate	Female urethra
Kupffer cell	Stellate macrophage	Liver	Spence tail	Axillary tail	Breast
Langerhans cell	Epidermal dendritic cell	Skin	Stensen duct	Parotid duct	Mouth
Langerhans cell	Centroacinar cell	Pancreas	Sylvius aqueduct	Cerebral aqueduct	Brain
Langerhans islet	Pancreatic islet	Pancreas	Sylvius fissure	Lateral cerebral sulcus	Brain
Leydig cell	Interstitial cell	Testis	Volkmann canal	Perforating canal; transverse canal	Bone tissue
Lieberkühn crypt	Intestinal gland	Small intestine	Wernicke area	Auditory association area	Brain
Meibomian gland	Tarsal gland	Eyelid			
Meissner corpuscle	Tactile corpuscle	Skin	Wharton duct	Submandibular duct	Mouth
Meissner plexus	Submucosal plexus	Intestinal wall	Willis circle	Cerebral arterial circle	Brain
Merkel cell	Tactile epithelial cell	Skin	Wirsung duct	Pancreatic duct	Pancreas
Montgomery gland	Areolar gland	Breast	Wolff(ian) duct	Mesonephric duct	Urogenital region (embryo)

TABLE 6   Scientific Symbols, Acronyms, and Abbreviations*

Hint▶ An abbreviation is a shortened form of a term, and an acronym is a type of abbreviation made from the first letter of each word (or word part) of a term. Acronyms are not standardized and thus may vary in usage. Some acronyms can represent more than one term, so check the context. If you cannot find an acronym you need in this list, check Tables 7 and 8.

ALERT! Items marked with ♦ are subject to health-threatening errors of interpretation and should no longer be used in *clinical* settings even though they are still often used in nonclinical or research settings. Items with ♦♦ are banned from use in clinical settings by The Joint Commission.

SYMBOL	TERM	SYMBOL	TERM	SYMBOL	TERM
1,25-D$_3$	1,25-dihydroxycholecalciferol	BMR	basal metabolic rate	DHEA	dehydroepiandrosterone
5-HT	serotonin	BP	blood pressure	DIP	distal interphalangeal (joint)
α-MSH	alpha melanocyte-stimulating hormone	BTB	blood-testis barrier	dl or dL	deciliter
Å	angstrom	°C	degrees Celsius or centigrade	DNA	deoxyribonucleic acid
AAA	American Association of Anatomists	C	calcitonin-producing (cell)	DRG	dorsal respiratory group
		C	Calorie or kilocalorie	dsRNA	double-stranded ribonucleic acid
ABP	androgen-binding protein	C†	cervical	E$_2$	estradiol
ACF	anterior cranial fossa	C†	complement (protein)	EAH	exercise-associated hyponatremia
Ach	acetylcholine	C or Cx	coccygeal	ECF	extracellular fluid
ACL	anterior cruciate ligament	CA	carbonic anhydrase	ECM	extracellular matrix
ACTH	adrenocorticotropic hormone	cal	calorie	EDV	end-diastolic volume
ADH	antidiuretic hormone	CART	cocaine- and amphetamine-regulated transcript	EF	ejection fraction
AFM	atomic force microscopy	cc♦	cubic centimeter	EFP	effective filtration pressure
AFP	alpha-fetoprotein	CCK	cholecystokinin	EGF	epidermal growth factor
AHF	antihemophilic factor	CD	collecting duct	EM	electron microscopy
AHG	antihemophilic globulin	CDK	cyclin-dependent kinase	ENS	enteric nervous system
AI	adequate intake	CD-*X*	cluster differentiation (blood cell marker system, where *X* is the cluster number)	EOP	endogenous opioid peptide
AIIS	anterior inferior iliac spine			Epi	epinephrine
ANH	atrial natriuretic hormone			EPOC	excess postexercise oxygen consumption
ANP	atrial natriuretic peptide	CN	cranial nerve		
ANS	autonomic nervous system	CNS	central nervous system	EPSP	excitatory postsynaptic potential
APC	antigen-presenting cell	CO	cardiac output	ER	endoplasmic reticulum
APS	American Physiological Society	CoA	coenzyme A	ERV	expiratory reserve volume
ASIS	anterior superior iliac spine	COMT	catechol-O-methyl transferase	ET	endothelin
AV	atrioventricular	COX	cyclooxygenase	ET-1	endothelin-1
AVP	arginine vasopressin (antidiuretic hormone)	CRH	corticotropin-releasing hormone	ETC	electron transport chain
		CT	calcitonin	ETS	electron transport system
B	bursa-equivalent tissue (as in B cell or B lymphocyte)	CT	chromosome territory	°F	degrees Fahrenheit
		CTFR	cystic fibrosis transmembrane conductance regulator	FA-1	fertilization antigen 1
BBB	blood-brain barrier	CTL	cytotoxic T lymphocyte	FASEB	Federation of American Societies for Experimental Biology
BCOP	blood colloid osmotic pressure	DA	dopamine		
BEAM	brain electrical activity map	DC	dendritic cell	FEV$_X$	forced expiratory volume (where *X* is the number of seconds)
BHP	blood hydrostatic pressure	DCT	distal convoluted tubule		
BM	basement membrane	DEJ	dermoepidermal junction (dermal-epidermal junction)	FCAT	Federative (International) Committee on Anatomical Terminology (see FICAT)
BMD	bone mineral density				
BMI	body mass index				

*These terms are given for information only and are not intended for use in clinical practice. Please consult policies at individual organizations for acceptable usage.

†Symbol is followed by a numeral.

Continued

T A B L E 6   **Scientific Symbols, Acronyms, and Abbreviations—cont'd**

SYMBOL	TERM	SYMBOL	TERM	SYMBOL	TERM
FICAT	Federative International Committee on Anatomical Terminology	HGP	human genome project	LP factor	leukocytosis-promoting factor
		H-K pump	hydrogen-potassium pump	μm	micrometer; micron
FM	fluorescence microscopy	HLA	human leukocyte antigen	M	mitotic phase of cell division
FRC	functional residual capacity	HPA	hypothalamus-pituitary-adrenal axis	M	mole or molar concentration
FSH	follicle-stimulating hormone			M	muscarinic
FVC	forced vital capacity	hPL	human placental lactogen	MAC	membrane attack complex
G_0	nondividing phase of cell life cycle	HR	heart rate	MALT	mucosal-associated lymphoid tissue
		Hz	Hertz (waves per second)		
G_1	first growth (or gap) phase of cell division	IC	inspiratory capacity	MAO	monoamine oxidase
		ICF	intracellular fluid	MC	mineralocorticoid
G_2	second growth (or gap) phase of cell division	IF	interferon	MCF	middle cranial fossa
		IFAA	International Federation of Associations of Anatomists	mDNA	mitochondrial deoxyribonucleic acid
GABA	gamma-aminobutyric acid (γ-aminobutyric acid)	IFCOP	interstitial fluid colliod osmotic pressure	Mgb	myoglobin
GAL	galanin			MHC	major histocompatibility complex
GAS	general adaptation syndrome	IFHP	interstitial fluid hydrostatic pressure	MMC	migrating motor complex
GC	glucocorticoid			mm Hg	millimeters of mercury (unit of pressure)
GFR	glomerular filtration rate	IFN	interferon		
GH	growth hormone	IFN-α	interferon alpha (fibroblast interferon)	MP	metocarpophalangeal (joint)
GHIH	growth hormone-inhibiting hormone (somatostatin)	IFN-β	interferon beta (leukocyte interferon)	mRNA	messenger ribonucleic acid
				MSH	melanocyte-stimulating hormone
GHRH	growth hormone-releasing hormone	IFN-γ	interferon gamma (immune interferon)	mtDNA	mitochondrial deoxyribonucleic acid
GHRL	ghrelin	IGF-1	insulin-like growth factor 1	mV	millivolt
GI	gastrointestinal	IL	interleukin	N	nicotinic
GIP	glucose-dependent insulinotropic peptide (gastric inhibitory peptide)	IL-X	interleukin (where X is a numeral [type])	NANC	nonadrenergic-noncholinergic
				NE	norepinephrine
GLP-1	glucagon-like peptide 1	IMP	integral membrane protein	Ngb	neuroglobin
Glu	glutamic acid, glutamate	IPSP	inhibitory postsynaptic potential	NK	natural killer (cell)
Gly	glycine	IRV	inspiratory reserve volume	nm	nanometer
GnRH	gonadotropin-releasing hormone	IS	immunological synapse	NMJ	neuromuscular junction
GPCR	G-protein-coupled receptor	IU◆◆	international unit	NPC	nuclear pore complex
HAPS	Human Anatomy and Physiology Society	JG	juxtaglomerular	NPY	neuropeptide Y
		°K	degrees Kelvin	NR	norepinephrine
$HbCO_2$	carbaminohemoglobin	L†	lumbar	OT	oxytocin
hCG	human chorionic gonadotropin	LDL	low-density lipoprotein	OXM	oxyntomodulin
HDL	high-density lipoprotein	LES	lower esophageal sphincter	P	pressure
hGH	human growth hormone	LH	luteinizing hormone	PA	alveolar pressure
HGH	human growth hormone	LM	light microscopy		

TABLE 6   Scientific Symbols, Acronyms, and Abbreviations—cont'd

SYMBOL	TERM	SYMBOL	TERM	SYMBOL	TERM
PB	barometric (atmospheric) pressure	RMP	resting membrane potential	Te or TE	*Terminologia Embryologica*
		RNA	ribonucleic acid	TEF	thermic effect of food
PCF	posterior cranial fossa	RNAi	RNA interference	TEM	transmission electron microscopy
pCi	picocurie	ROM	range of motion	TH	thyroid hormone
PCL	posterior cruciate ligament	rRNA	ribosomal ribonucleic acid	Th or TH	*Terminologia Histologica*
PCM	pericentriolar material	RV	residual volume	TLC	total lung capacity
PCO_2	(partial) pressure of carbon dioxide	S	solubility	TLR	toll-like receptor
		$S^\dagger$	sacral	Tm	transport maximum
PCT	proximal convoluted tubule	SA	sinoatrial	Tmax	transport maximum
PG	prostaglandin	SEM	scanning electron microscopy	TMR	total metabolic rate
PIH	prolactin-inhibiting hormone	SER	smooth endoplasmic reticulum	TNF	tumor necrosis factor
PIIS	posterior inferior iliac spine	SFO	subfornical organ	TP	transverse process
Pip	intrapleural pressure	SITS	supraspinatus, infraspinatus, teres minor, subscapularis (muscles)	t-PA	tissue plasminogen activator
PIP	proximal interphalangeal (joint)			TPA	tissue plasminogen activator
PNS	peripheral nervous system			TPR	total peripheral resistance
PO_2	(partial) pressure of oxygen	snRNA	small nuclear ribonucleic acid	TRH	thyrotropin-releasing hormone
PP	pancreatic polypeptide	snRNP	small nuclear ribonucleic particle; small nuclear ribonucleoprotein	tRNA	transfer ribonucleic acid
PRH	prolactin-releasing hormone			TSH	thyroid-stimulating hormone
PRL	prolactin	SNS	somatic nervous system	TV	tidal volume
PSA	prostate-specific antigen	SP	spinous process	TX	thromboxane
PSIS	posterior superior iliac spine	SPCA	serum prothrombin conversion accelerator	UES	upper esophageal sphincter
PTA	plasma thromboplastin antecedent			UV	ultraviolet
		SR	sarcoplasmic reticulum	UV-A	A range of ultraviolet
PTC	plasma thromboplastin component	SS	somatostatin	UVA	A range of ultraviolet
		STH	somatotropin; somatotropic hormone	UV-B	B range of ultraviolet
PTH	parathyroid hormone			UVB	B range of ultraviolet
PX	pressure (where X is the type of pressure)	SV	stroke volume	UV-C	C range of ultraviolet
		SWS	slow-wave sleep	UVC	C range of ultraviolet
$PYY_{3\text{-}36}$	peptide $Y_{3\text{-}36}$	t	thickness	V	volt or voltage
RAAS	renin-angiotensin-aldosterone system	T	thymus (as in T cell or T lymphocyte)	VC	vital capacity
RAS	reticular activating system	$T^\dagger$	thoracic	VIP	vasoactive intestinal peptide
RBC	red blood cell	T_3	triiodothyronine	$VO_{2\,max}$	maximum oxygen consumption
RDA	recommended dietary (or daily) allowance	T_4	tetraiodothyronine	VNO	vomeronasal organ
		Ta or TA	*Terminologia Anatomica*	VRG	ventral respiratory group
REM	rapid-eye movement [sleep]	TAL	thick ascending limb	WBC	white blood cell
RER	rough endoplasmic reticulum	tALH	thin ascending limb (of Henle)	X	larger of two sex chromosomes
Rh	Rhesus (blood antigen)	TBG	thyroid-binding globulin	Y	smaller of two sex chromosomes
RISC	RNA-induced silencing complex	TCA	tricarboxylic acid (cycle)		

T A B L E 7   Medical Symbols, Acronyms, and Abbreviations*

An abbreviation is a shortened form of a term, and an acronym is a type of abbreviation made from the first letter of each word (or word part) of a term. Acronyms are not standardized and thus may vary in usage, especially in different specialized fields or specialties. Some acronyms can represent more than one term, so check the context. If you cannot find an acronym you need in this list, check Tables 6 and 8.

ALERT! Items marked with ♦ are subject to health-threatening errors of interpretation and should no longer be used in *clinical* settings even though they are still often used in nonclinical or research settings. Items with ♦♦ are banned from use in clinical settings by The Joint Commission.

SYMBOL	TERM	SYMBOL	TERM	SYMBOL	TERM
aa	of each	b.i.d.	twice a day	CML	chronic myeloid leukemia
a̅.c̅.	before meals	b.m.	bowel movement	CNM	certified nurse midwife
ABC	aspiration biopsy cytology	BMI	body mass index	CNS	central nervous system
AD	Alzheimer disease	BMR	basal metabolic rate	CNS	clinical nurse specialist
ad lib.	as much as desired	BP	blood pressure	Co	cobalt
AED	automatic external defibrillator	BPH	benign prostatic hypertrophy	COPD	chronic obstructive pulmonary disease (or disorder)
AF or A-fib	atrial fibrillation	BPPV	benign paroxysmal positional vertigo	CP	cerebral palsy
AI	adequate intake	BRP	bathroom privileges	CPAP	continuous positive airway pressure
AICD	automatic internal cardiac defibrillator	BSE	bovine spongiform encephalopathy	CPR	cardiopulmonary resuscitation
AIDS	acquired immune deficiency syndrome	BUN	blood urea nitrogen	CRNA	certified registered nurse anesthetist
AK	astigmatic keratotomy	c̅	with	CRT	certified respiratory therapist
alb.	albumin	CA-125	cancer antigen 125	CSR	Cheyne-Stokes respiration
ALK	automated lamellar keratoplasty	CAD	coronary artery disease	CT	computed tomography
ALL	acute lymphocytic leukemia	CAPD	continuous ambulatory peritoneal dialysis	CVA	cerebrovascular accident, stroke
am	before noon	CBC	complete blood cell count	CVS	chorionic villus sampling
AMA	American Medical Association	CCU	coronary care unit	DA	dental assistant
AML	acute myeloid leukemia	CDA	certified dental assistant	DC	doctor of chiropractic
amt.	amount	CDC	Centers for Disease Control and Prevention	D & C	dilation and curettage
ante	before	CDN	certified dialysis nurse	d/c♦	discontinue
aq.	water	CEN	certified emergency nurse	DDS	doctor of dental surgery
ARDS	acute respiratory distress syndrome	CF	cystic fibrosis	DJD	degenerative joint disease (osteoarthritis)
ART	assistive reproductive technique (or technology)	CFTR	cystic fibrosis transmembrane conductance regulator	DM	diabetes mellitus
ASA	acetylsalicylic acid (aspirin)	CHF	congestive heart failure	DMD	doctor of dental medicine
ASCP	American Society for Clinical Pathology	CICU	coronary intensive care unit	DMD	Duchenne muscular dystrophy
AV.	average	CJD	see vCJD	DMT	doctor of medical technology
AZT♦	azidothymidine	CK	conductive keratoplasty	DN	doctor of nursing
Ba	barium	CLL	chronic lymphocytic leukemia	DNS	doctor of nursing science
BE	barium enema	CLT	certified laboratory technician	DO	doctor of optometry
		CMA	certified medical assistant	DO	doctor of osteopathy
				DOA	dead on arrival

*These terms are given for information only and are not intended for use in clinical practice. Please consult policies at individual organizations for acceptable usage.

TABLE 7 **Medical Symbols, Acronyms, and Abbreviations—cont'd**

SYMBOL	TERM	SYMBOL	TERM	SYMBOL	TERM
DON	director of nursing	GNP	gerontological nurse practitioner	MD	medical doctor
DP	doctor of pharmacy	GP	general practitioner	MED	male erectile dysfunction
DP	doctor of podiatry	GU	genitourinary	MEG	magnetoencephalography
DPM	doctor of podiatric medicine	h.	hour	MH	malignant hyperthermia
DUB	dysfunctional uterine bleeding	HAART	highly active antiretroviral therapy	MI	myocardial infarction
DVT	deep vein thrombosis	HDL	high-density lipoprotein	MRI	magnetic resonance imaging
Dx	diagnosis	h.s.	at bedtime	MVP	mitral valve prolapse
DXA	dual-energy x-ray absorptiometry	H_2O	water	NAT	nucleic acid test
ECG	electrocardiogram	Hb	hemoglobin	NCT	nuclear cardiology technologist
ED	erectile dysfunction	HCT or Hct	hematocrit	ND	doctor of naturopathy
EDC	expected date of confinement			ND	doctor of nursing
EEG	electroencephalogram; electroencephalograph	HD	Huntington disease	NFP	natural family planning
		HEC	human-engineered chromosome	NICU	neonatal intensive care unit
EENT	ear, eye, nose, throat	Hib	*Haemophilus influenzae* type B	NIH	National Institutes of Health
EKG	electrocardiogram	HIV	human immunodeficiency virus	NMR	nuclear magnetic resonance (imaging)
EMT	emergency medical technician	HMD	hyaline membrane disease		
EMT-P	emergency medical technician paramedic	HTN	hypertension	NMT	nuclear medicine technologist
		ICU	intensive care unit	non rep.	do not repeat
ENT	ear nose throat (specialist)	ID	intradermal	NP	nurse practitioner
EP	evoked potential	IM	intramuscular	NPO	nothing by mouth
ER	emergency room	IPPB	intermittent positive pressure breathing	NREMT-P	National Registry emergency medical technician paramedic
ERT	estrogen replacement therapy				
ESR	erythrocyte sedimentation rate	IRDS	infant respiratory distress syndrome	NRI	norepinephrine reuptake inhibitor
FA-1	fertilization antigen 1	ISMP	Institute for Safe Medication Practices	NSAID	nonsteroidal antiinflammatory drug
FAS	fetal alcohol syndrome			OCN	oncology certified nurse
FDA	Food and Drug Administration	IVF	in vitro fertilization	OD	doctor of optometry
FED	female erectile dysfunction	JRA	juvenile rheumatoid arthritis	OR	operating room
FM	fibromyalgia	KUB	kidney, ureter, and bladder	ORT	oral rehydration therapy
fMRI	functional magnetic resonance imaging	LASIK	laser-assisted in situ keratomileusis	OT	occupational therapist
				OTR	registered occupational therapist
FNP	family nurse practitioner	LDL	low-density lipoprotein	PA	physician assistant
FSX	fragile X syndrome	LGI	lower gastrointestinal (study)	PAC	premature atrial contraction
FUO	fever of undetermined origin	LPN	licensed practical nurse	PAD	peripheral arterial disease
GAS	general adaptation syndrome	LRI	lower respiratory infection	Pap	Papanicolaou
GDM	gestational diabetes mellitus	LVN	licensed vocational nurse	p.c.	after meals
GERD	gastroesophageal reflux disease (or disorder)	LVRS	lung volume reduction surgery	PCI	percutaneous coronary intervention
		MAb	monoclonal antibody	PCM	protein-calorie malnutrition
GI	gastrointestinal	MAOI	monoamine oxidase inhibitor		

Continued

T A B L E 7   Medical Symbols, Acronyms, and Abbreviations—cont'd

SYMBOL	TERM
PCOD	polycystic ovary disease
PCOS	polycystic ovary syndrome
PCP	primary care physician
PD	doctor of pharmacy
PD	Parkinson disease
PDT	photodynamic therapy
per	by
PET	positron emission tomography
PH	past history
PharmD	doctor of pharmacy
PhD	doctor of philosophy
PI	previous illness
PID	pelvic inflammatory disease
PIH	pregnancy-induced hypertension
PKD	polycystic kidney disease
PKU	phenylketonuria
pm	after noon
PMS	premenstrual syndrome
PNP	pediatric nurse practitioner
PP	pulse pressure
PRK	photorefractive keratotomy
p.r.n.	as needed
PSA	prostate-specific antigen
PT	physical therapist
PVC	premature ventricular contraction
PVD	peripheral vascular disease
q.	every
q.d.♦♦	every day
q.h.	every hour
q.i.d.	four times a day
q.n.s.	quantity not sufficient
q.o.d.♦♦	every other day
q.s.	quantity required or sufficient
RA	radiographic absorptiometry
RA	rheumatoid arthritis

SYMBOL	TERM
RBC	red blood cell
RDA	recommended dietary (or daily) allowance
RDA	registered dental assistant
RGB	Roux-en-Y gastric bypass
RK	radial keratotomy
RN	registered nurse
RNA	registered nurse anesthetist
RNA	registered nursing assistant
RPFT	registered pulmonary function technologist
RPh	registered pharmacist
RPT	registered physical therapist
RPT	registered physiotherapist
RRA	registered record(s) administrator
RRT	registered respiratory therapist
RT	radiological technologist
RT	respiratory therapist
RVT	registered vascular technologist
R	prescription
$\bar{s}$	without
SAD	seasonal affective disorder
SARS	severe acute respiratory syndrome
SCID	severe combined immunodeficiency
SIDS	sudden infant death syndrome
SIRS	systemic inflammatory response syndrome
SK	streptokinase
SLE	systemic lupus erythematosus
SLN	sentinel lymph node
SNRI	serotonin and norepinephrine reuptake inhibitor
SPECT	single-photon emission computed tomography
sp. gr.	specific gravity

SYMBOL	TERM
SQ♦	subcutaneous
$\bar{s}s$♦	sliding scale or one half
SSRI♦	serotonin-specific reuptake inhibitor
stat.	at once, immediately
STD	sexually transmitted disease
STI	sexually transmitted infection
SW	social worker
T_3	triiodothyronine
T_4	tetraiodothyronine
T & A	tonsillectomy and adenoidectomy
T.B.	tuberculosis
TENS	transcutaneous electrical nerve stimulation
THR	total hip replacement
t.i.d.	three times a day
TJC	The Joint Commission
TMR	transmyocardial laser revascularization
t-PA	tissue plasminogen activator
TPA	tissue plasminogen activator
TPR	temperature, pulse, respiration
TS	Turner syndrome
TSD	Tay-Sachs disease
TSS	toxic shock syndrome
TUR	transurethral resection
UGI	upper gastrointestinal
URI	upper respiratory infection
UTI	urinary tract infection
vCJD	variant Creutzfeldt-Jakob disease
VF or V-fib	ventricular fibrillation
VZV	varicella zoster virus
WBC	white blood cell
XP	xeroderma pigmentosum

TABLE 8 Chemical Symbols, Formulas, and Acronyms*

 A chemical symbol is an abbreviation of an element. A chemical formula is a term made of chemical symbols in a way that usually represents the kind of elements present in a molecule and the number of atoms of each of those elements. This table lists symbols of all the known elements and the formulae of many essential molecules. Some acronyms or other type of abbreviations of chemical names are also provided. If you cannot find what you need in this list, check Tables 6 and 7.

ALERT! Items marked with ♦ are subject to health-threatening errors of interpretation and should no longer be used in *clinical* settings even though they are still often used in nonclinical or research settings. Items with ♦♦ are banned from use in clinical settings by The Joint Commission.

SYMBOL	TERM	SYMBOL	TERM	SYMBOL	TERM
−	negative	$C_6H_{12}O_6$	glucose; dextrose	F	phenylalanine
+	positive	Ca	calcium	FAD	flavin adenine dinucleotide
1,25-D$_3$	1,25-dihydroxycholecalciferol	Ca^{++}	calcium ion	Fe	iron
5-HT	serotonin	CaCl$_2$	calcium chloride	FFA	free fatty acid
A	alanine	CaCO$_3$	calcium carbonate	Fm	fermium
A or a	adenine	Ca$_3$(PO$_4$)$_2$	calcium phosphate	Fr	francium
Ac	actinium	cAMP	cyclic adenosine monophosphate	G	glycine
ACh or Ach	acetylcholine	CCK	cholecystokinin	G or g	guanine
ADA	adenosine deaminase	Cd	cadmium	Ga	gallium
ADP	adenosine diphosphate	Ce	cerium	GABA	gamma-aminobutyric acid
Ag	silver	Cf	californium	Gd	gadolinium
Al	aluminum	Cl	chlorine	Ge	germanium
Ala	alanine	Cl$^-$	chloride ion	Gln	glutamine
Am	americium	Cm	curium	Glu	glutamic acid
Ar	argon	Cn	copernicium	Gly	glycine
Arg	arginine	Co	cobalt	^{2}H	deuterium
As	arsenic	CoQ$_{10}$	coenzyme Q$_{10}$	^{3}H	tritium
Asn	asparagine	CP	creatine phosphate	H	hydrogen
Asp	aspartic acid	Cr	chromium	H$^+$	hydrogen ion; proton
At	astatine	Cs	cesium	H$_2$CO$_3$	carbonic acid
ATP	adenosine triphosphate	Cu	copper	H$_2$O	water
Au	gold	Cys	cysteine	Hb	hemoglobin
AZT	azidothymidine	D	aspartic acid	HbCO$_2$	carbaminohemoglobin
B	boron	DA	dopamine	HbO$_2$	oxyhemoglobin
Ba	barium	Db	dubnium	HCl♦	hydrochloric acid
BB	base bicarbonate	DNA	deoxyribonucleic acid	HCO$_3^-$	bicarbonate ion
Be	beryllium	Ds	darmstadtium	HDL	high-density lipoprotein
Bh	bohrium	dsRNA	double-stranded ribonucleic acid	He	helium
Bi	bismuth	Dy	dysprosium	Hf	hafnium
Bk	berkelium	e$^-$	electron	Hg	mercury
Br	bromine	E	glutamic acid	His	histidine
^{13}C	carbon-13	Epi	epinephrine	Ho	holmium
^{14}C	carbon-14	Epo or EPO	erythropoietin	HPO$_4^-$	hydrogen phosphate ion
C	carbon	Er	erbium	Hs	hassium
C	complement (protein)	Es	einsteinium	I	iodine
C	cysteine	Eu	europium	I$^-$	iodine ion
C or c	cytosine	F	fluorine	I or Ile	isoleucine

*These terms are given for information only and are not intended for use in clinical practice. Please consult policies at individual organizations for acceptable usage.

Continued

TABLE 8   **Chemical Symbols, Formulas, and Acronyms—cont'd**

SYMBOL	TERM	SYMBOL	TERM	SYMBOL	TERM
In	indium	Nd	neodymium	Rn	radon
Ir	iridium	Ne	neon	RNA	ribonucleic acid
K	lysine	NE	norepinephrine	RNA	ribosomal ribonucleic acid
K	potassium	Ngb	neuroglobin	Ru	ruthenium
K^+	potassium ion	NH_2	amine group	S	serine
KCl	potassium chloride	NH_3	ammonia	S	sulfur
Kr	krypton	Ni	nickel	Sb	antimony
L	leucine	No	nobelium	Sc	scandium
La	lanthanum	NO	nitric oxide	Se	selenium
LDL	low-density lipoprotein	Np	neptunium	Sec	selenocysteine
Leu	leucine	NPY	neuropeptide Y	Ser	serine
Li	lithium	NS	nucleoside	Sg	seaborgium
Lr	lawrencium	NT	nucleotide	Si	silicon
Lu	lutetium	O	oxygen	siRNA	short interfering ribonucleic acid; also called short interfering ribonucleic acid or silencing ribonucleic acid
Lys	lysine	O_2	oxygen molecule		
M	methionine	OH^-	hydroxide ion		
Md	mendelevium	Os	osmium		
mDNA	mitochondrial deoxyribonucleic acid	P	phosphorus; phosphate group	Sm	samarium
		P	proline	Sn	tin
Met	methionine	p^+	proton	snRNA	small nuclear ribonucleic acid
Mg	magnesium	Pa	rotactinium	snRNP	small nuclear ribonucleoprotein
Mg^{++}	magnesium ion	Pb	lead	SO_4^-	sulfate ion
$MgCl_2$	magnesium chloride	Pd	palladium	Sr^-	strontium
Mn	manganese	PG	prostaglandin	T	threonine
Mo	molybdenum	Phe	phenylalanine	T or t	thymine
mRNA	messenger ribonucleic acid	Pi	inorganic phosphate	T_3	triiodothyronine
Mt	meitnerium	Pm	promethium	T_4	tetraiodothyronine
mtDNA	mitochondrial deoxyribonucleic acid	Po	polonium	Ta	tantalum
		PO_4^-	phosphate ion	Tb	terbium
N	asparagine	Pr^-	praseodymium	Tc	technetium
N	nitrogen	Pro^-	proline	TCA	tricarboxylic acid
n^0	neutron	Pt	platinum	Te	tellurium
N_2	nitrogen molecule	Pu	plutonium	Th	thorium
Na	sodium	Q	glutamine	Thr	threonine
Na^+	sodium ion	R	arginine	Ti	titanium
NaCl	sodium chloride; table salt	Ra	radium	Tl	thallium
NAD	nicotinamide adenine dinucleotide	Rb	rubidium	Tm	thulium
		Re	rhenium	tRNA	transfer ribonucleic acid
$NaHCO_3$	sodium bicarbonate	Rf	rutherfordium	Trp	tryptophan
Na_2SO_4	sodium sulfate	Rg	roentgenium	Tyr	tyrosine
NAG	N-acetylglucosamine	Rh	rhodium	U	uranium
Nb	niobium			U	selenocysteine

TABLE 8 **Chemical Symbols, Formulas, and Acronyms—cont'd**

SYMBOL	TERM	SYMBOL	TERM	SYMBOL	TERM
U or u	uracil	Uut	ununtrium	W	tungsten
Uub	ununbium	Uuu	ununium	Xe	xenon
Uuh	ununhexium	V	valine	Y	tyrosine
Uuo	ununoctium	V	vanadium	Y	yttrium
Uup	ununpentium	Val	valine	Yb	ytterbium
Uuq	ununquadium	VIP	vasoactive intestinal peptide	Zn	zinc
Uus	ununseptium	W	tryptophan	Zr	zirconium

TABLE 9 **Greek Alphabet**

 Although letters of the Roman alphabet (the letters you are reading now) are often used to designate different categories of things, scientific and medical terminology sometimes uses letters from the Greek alphabet. You may encounter either lower or upper case Greek letters, or both, in scientific terms. Use this list to look up Greek letters that you run across in your study of anatomy and physiology.

UPPERCASE	LOWERCASE	NAME OF LETTER
Α	α	alpha
Β	β	beta
Γ	γ	gamma
Δ	δ	delta
Ε	ε	epsilon
Ζ	ζ	zeta
Η	η	eta
Θ	θ	theta
Ι	ι	iota
Κ	κ	kappa
Λ	λ	lambda
Μ	μ	mu
Ν	ν	nu
Ξ	ξ	xi
Ο	ο	omicron
Π	π	pi
Ρ	ρ	rho
Σ	σ	sigma
Τ	τ	tau
Υ	υ	upsilon
Φ	φ	phi
Χ	χ	chi
Ψ	ψ	psi
Ω	ω	omega

TABLE 10 **Roman Numerals**

 Roman Numerals Science terms often incorporate arabic numerals (1, 2, 3, etc.) but occasionally use Roman numerals. In the Roman system, the letters I, V, X, L, C, D, and M are used as numerals. The list that follows the table of basic symbols shows how the symbols are used to form numerals. This is a brief list of some Roman numerals and their arabic equivalents that are often used in scientific expressions. (Note: The Roman system has no symbol for zero; if needed, an arabic 0 may be used in a term.)

ROMAN NUMERAL	ARABIC NUMERAL	ROMAN NUMERAL	ARABIC NUMERAL
I	**1**	XVI	16
II	2	XVII	17
III	3	XVIII	18
IV	4	XIX	19
V	**5**	XX	20
VI	6	XXX	30
VII	7	XL	40
VIII	8	**L**	**50**
IX	9	LX	60
X	**10**	LXX	70
XI	11	LXXX	80
XII	12	XC	90
XIII	13	**C**	**100**
XIV	14	**D**	**500**
XV	15	**M**	**1000**

Boldface numbers represent basic Roman symbols and their arabic equivalents.